Y0-DJN-982

ISA Guide To Measurement Conversions

George Platt

Instrument Society of America

Printed in the United States of America

INSTRUMENTS SOCIETY OF AMERICA
67 Alexander Drive
P. O. Box 12277
Research Triangle Park
North Carolina 27709

Library of Congress Cataloging-in-Publication Data

Platt, George, 1920-
ISA Guide to Measurement conversions / by George Platt.
p. cm.
Includes index.
ISBN 1-55617-489-6
1. Metric system—Conversion tables. I. Title.
QC94.P55 193 — 93-27095
530.8'1—dc20 — CIP

Great care has been taken to provide reliable data for converting measurement units, and the information is believed to be correct. However, the author and the publisher cannot be held responsible for discrepancies or errors, if any, or for the consequences of their use.

In some cases, different authorities provide data that differ. Differences may arise also when conversion values are derived by calculations.

The conversion listings do not imply that the use of any particular measuring unit is recommended or meets current practices of usage.

TABLE OF CONTENTS

Tables

Appendix

Tables

Appendix

CHAPTER 1

PREFIXES AND SYMBOLS

1.1 The International System of Units

The International System of Units (SI), established in 1960, is an improved form of the worldwide metric system of measurement. "SI," the French acronym for this improved system, is used in all languages. The United States, which has traditionally used the English system of measuring units — foot, pound, etc. — has endorsed the SI system — meter, kilogram, etc. This book lists conversions for SI units and traditional units.

The plain measuring units of the SI system and its forerunner are often too large or too small for convenient use because some measurement numbers would require very many zeroes to the left or the right of a decimal point. However, all the units can be made larger or smaller by adding prefixes. Thus, a kilometer (in speech, all the prefixes accent the first syllable) equals a thousand meters, a centimeter equals one hundredth of a meter. The prefixes and their values, in three different notations, are listed in Table 1. For examples of the utility of the prefixes, see Chapter 6.

Table 1
SI UNIT MODIFIERS

Prefix	Modern	Exponential	Numerical
yotta*	E+24	10^{24}	1 000 000 000 000 000 000 000 000
zetta*	E+21	10^{21}	1 000 000 000 000 000 000 000
exa	E+18	10^{18}	1 000 000 000 000 000 000
peta	E+15	10^{15}	1 000 000 000 000 000
tera	E+12	10^{12}	1 000 000 000 000
giga	E+09	10^{9}	1 000 000 000
mega	E+06	10^{6}	1 000 000
myria**	E+04	10^{4}	10 000
kilo	E+03	10^{3}	1 000
hecto	E+02	10^{2}	100
deka	E+01	10^{1}	10
- - ***	E+00	10^{0}	1
deci	E-01	10^{-1}	0.1
centi	E-02	10^{-2}	0.01
milli	E-03	10^{-3}	0.001
micro	E-06	10^{-6}	0.000 001
nano	E-09	10^{-9}	0.000 000 001
pico	E-12	10^{-12}	0.000 000 000 001
femto	E-15	10^{-15}	0.000 000 000 000 001
atto	E-18	10^{-18}	0.000 000 000 000 000 001
zepto*	E-21	10^{-21}	0.000 000 000 000 000 000 001
yocto*	E-24	10^{-24}	0.000 000 000 000 000 000 000 001

* Prefixes *yetta*, *zetta*, *zepto*, and *yocto* are listed by the British.
** *myria* is obsolete.
*** There is no prefix for E+00.

In this book, numbers that have more than five digits before a decimal point or more than two zeroes immediately after the decimal point are presented in the *E* notation of Table 1. This notation is used for electronic data transmission and computers, and avoids the awkward superscripts of the traditional exponential notation using powers of 10.

1.2 Symbols and Abbreviations

Symbol	*Meaning*
^	a value is exact, i.e., as though the last digit shown were followed by an infinite number of zeroes
5/9^ (typical)	the entire fraction is exact
x	multiply
x 1	inexact equality (see Chapter 2, By Precisely)
x 1^	exact equality (see Chapter 2, By Precisely)
/	divide
±	plus or minus
@	at
AGA	American Gas Association
approx.	approximate
Brit.	British
C	Celsius (formerly centigrade)
Can.	Canadian
dB	decibel
deg.	degree
e.g.	for example
EMU	electromagnetic centimeter-gram-second unit
Eng.	English
ESU	electrostatic centimeter-gram-second unit
etc.	and so forth
F	Fahrenheit
hr	hour
intl.	international
IT	International Table
ITS	International Temperature Scale
k	kelvin
ln	natural logarithm
log	common logarithm
min	minute
no.	number
Queb.	Quebec
R	Rankine
sec	second
SI	International System of Units
temp.	temperature
US	United States
yr	year

CHAPTER 2

CONVERSION TABLES AND COLUMNS

The conversions of measuring units are presented in Table 2, an alphabetical listing, and in Table 3, a listing by measurement groups. Both tables have columns, which are described below.

Convert: This column lists the measuring units or terms that are to be converted to other units. Thus, "foot" could be converted to "inches" or "gallon" to "liters."

To Equivalent: This column lists the new units to be obtained by conversion. Thus, from the *Convert* column above, the equivalent new units are "inches" or "liters," respectively.

By Precisely:* This column lists the appropriate factor or factors for each conversion. Thus, a "grain per gallon" is converted to "pounds per gallon" with a conversion factor "/7 000^," indicating that one grain per gallon equals one pound per gallon divided by 7 000, or one 7 000th of a pound per gallon. (One pound equals 7 000 grains.) The caret symbol after the 7 000 signifies that the factor is exact. It may be more more convenient in some cases to use a multiplier rather than a divider for the conversion, so an alternative equivalent factor, "x 1.428 6 E-04," is also listed. This multiplier is very precise, having five significant digits (see Section 4.1), yet it is not exact.

Where there is more than one value for a given conversion, the values are stated in order of decreasing accuracy. The choice of factors is the user's.

A conversion factor "x 1^" indicates that the conversion is exact as shown. The two measuring units or terms involved are fully equal. (The caret symbol is here only to explain the conversion and is not expected to be adopted or shown in actual practice.)

A conversion factor "x 1" without the caret indicates that (1) the two units or terms are equal at least to the extent of the number of significant digits shown, and (2) the conversion is actually inexact or is not known to be exact. For example: One US statute foot legally equals 1.000 002 survey feet. The two kinds of "foot" are not exactly equal; their difference is 2 parts per million. Nevertheless, the statute foot and the survey foot are effectively equal for all but extremely specialized purposes and are so indicated by the "x 1" symbolism without a caret. Thus, a practical equivalency but not exactness is expressed.

Or Within ±5.0 %: Approximate conversion values may be useful for quick checking of detailed calculations, preliminary estimates, or mental calculations. This column lists approximate values that are accurate within the band of +5.0 % to -5.0 % of their precise value. Here, too, there may be conversion values stated in order of decreasing accuracy.

* This example follows the practices of the American National Standards Institute and recommended international practice regarding (a) the use of the *E* notation for powers of 10, and (b) the dividing of numbers into digit groups of three for ease of reading and to avoid possible confusion caused by foreign countries' usage of the period and commas for numbers versus the different usage by the United States.

CHAPTER 3

UNLISTED CONVERSIONS

It is not feasible to list conversions for all possible permutations of measuring units. But many unlisted conversions may nevertheless be obtained, as indicated below.

3.1 A Simple Conversion

How does one convert "mile" to "marathon"?

The Easy Way:

In general, this method has a 50 % probability of causing the value of the last valid digit of the converted number to be in error by 1. The resulting error may be acceptable in many cases.

Example: Tables 2 and 3 list "marathon to miles x 26.219," so 1 marathon = 26.219 miles. But the reverse, "mile" to "marathon," is not listed. For such a case, the conversion factor should be the reciprocal, or inverse, of the known conversion factor. The reciprocal is obtained by dividing the number 1 by the known factor:

$$1^\wedge \cancel{\text{mile}} \times \frac{1 \text{ marathon}}{26.219 \cancel{\text{miles}}} = 0.038\,140\,280 \text{ marathon}$$

Rounded to five significant digits, 1 mile = 0.038 140 marathon.

The Proper Way:

In principle, to achieve maximum accuracy in conversions, all the calculation numbers should first have more validly significant digits, so far as practical, than are required in the final answer (see Section 4.1); this should be the goal. The Easy Way, above, uses 26.219 in the calculation; the Proper Way is to use the more accurate number, 26.218 75, as shown below. In any event, the conversion calculation should be performed, and then the answer may be rounded as necessary or desired (see Section 4.3). The sequence — first calculate, then round — holds also for the Easy Way, above.

A reference states that a marathon distance is 26 miles, 385 yards, which equals 26.218 75 miles.

$$1^\wedge \cancel{\text{mile}} \times \frac{1 \text{ marathon}}{26.218\,75 \cancel{\text{miles}}} = 0.038\,140\,644 \text{ marathon}$$

Rounded to five significant digits, 1 mile = 0.038 141 marathon is a trifle more accurate than the previous result. The difference is 1 part in 38 141, or 0.002 6 %.

3.2 A Complex Conversion

Conversions are often complex. An example might be to convert a speed of 5 feet per year to millimeters per second. Using listed conversions:

$$5^\wedge\,\frac{\cancel{\text{feet}}}{\cancel{\text{year}}} \times \frac{0.304^\wedge\,\cancel{\text{meter}}}{\cancel{\text{foot}}} \times \frac{1\,000^\wedge\,\text{millimeters}}{\cancel{\text{meter}}} \times \frac{1\,\cancel{\text{year}}}{365^\wedge\,\cancel{\text{days}}}$$

$$\times\, 1\,\frac{\cancel{\text{day}}}{24^\wedge\,\cancel{\text{hours}}} \times \frac{1\,\cancel{\text{hour}}}{3\,600^\wedge\,\cancel{\text{seconds}}} = 4.832\,6\text{ E}-05\,\frac{\text{millimeters}}{\text{second}}$$

The calculated answer is 4.832 572 E-05 millimeters per second, or 4.832 6 millimeters per second when rounded to five significant digits (see Section 4.3). Note that all the measuring units cancel out except for those ("millimeters" and "second") that are the desired final units.

CHAPTER 4

SIGNIFICANT DIGITS

4.1 Definition

A *significant digit*, or *significant figure*, is any digit, including zero, that serves to define a value or quantity. Zeroes used only to show where a decimal point is placed in a number are not significant. Example: 6, 8, and 3 are the only significant digits in the numbers 683, 6.83, 0.006 83, and 6.83 E+09, but the four numbers have different magnitudes.

Another case is the number 68 300, for which the zeroes may or may not be significant depending on whether they exist (1) because they are part of the measurement or rather (2) because they serve merely to indicate the magnitude of the number. If both zeroes are not significant, the ambiguity can be resolved, if necessary, by writing the number as 6.83 E+04, or 6.83×10^4, or as 68.3 thousand. If the first zero is significant because it was determined by measurement but the second zero is not significant because it is used only to indicate magnitude, the unambiguous number would be 6.830 E+04 or 68.30 thousand. If both zeroes are significant, the unambiguous number would be 6.830 0 E+04 or 68.300 thousand. All zeroes following decimal digits are considered significant, and they may be appended only if they really are significant.

The more refined a measurement is, that is, the greater its precision, the greater is the number of significant digits that can be considered valid.

4.2 Major Rules for Calculations

4.2.1 Rationale

Numbers based on measurements are more or less precise depending on the quality of the measuring tools, how the tools are used, sensitivity to environmental changes, and other factors. Numbers based on estimates are also imprecise. Numbers and fractions whose values are defined exactly and numbers that are exact counts are perfectly precise or exact.

The precision of a calculated result -- the valid number of significant digits -- may be limited by an imprecise calculation number but is not limited by an exact number.

4.2.2 Addition and Subtraction

A calculated answer shall contain no significant digits farther to the right than exist in the calculation number whose rightmost significant digit is least far to the right. So the proper sum of 15.4 plus 1.327 is 16.7 because the 4 is less far to the right than is the 2.

4.2.3 Multiplication, Division, and Extraction of Roots

A calculated answer shall contain no more significant digits than exist in the calculation number having the fewest significant digits. So 2 436 divided by 512 calculates to 4.757 812 6, which is valid to only 4.76 because the three-digit 512 has fewer significant digits than does 2 436. Also the square root of 38.7 is 6.220 932 406, which is valid to only 6.22.

4.2.4 Logarithms

The characteristic denotes only magnitude and is not significant. The mantissa is significant and should contain the same number of digits as are in the corresponding number. Thus, the common logarithm of the number 103.2 should be taken as 2.013 7 rather than 2.013 679.

4.2.5 Errors

Error values need not carry more than two significant digits.

4.2.6 Values with Tolerances

For values having specific tolerances, as for machine work, a reference such as American National Standards Institute Standard ANSI/IEEE 268-1982 may be consulted.

4.3 Rounding Conversion Values

Especially when a calculator is used, conversion values for measurements are frequently calculated to more significant digits than are valid. The number of digits that are valid depends on the precision of the particular measurement values and other numbers involved in the calculation. So a calculated value may have to drop some digits in order to avoid giving a falsely good picture of the precision of the conversion. This requires proper rounding of the calculated value.

Rounding may also be useful when a totally valid conversion value or number has more significant digits than are needed and the person using the information is encumbered by unnecessary precision. (In an extreme case: The value of the number pi has been determined to over one million places; this precision is obviously not needed for determining the area of a circle.)

Rounding can be done as follows:

(1) Decide on the farthest-to-the-right significant digit that is to be retained.

(2) Look at that digit to be retained and note the digit immediately to its right, the first digit to be discarded.

(3) (a) If the first digit discarded is 0, 1, 2, 3, or 4, do not change the retained-digit value.

(b) If the first digit discarded is 5 followed by only zeroes and the last digit retained is *even*, do not change the retained-digit value.

If the first digit discarded is 5 followed by only zeroes and the last digit retained is *odd*, increase the retained-digit value by 1.

If the first digit discarded is 5 followed by at least one digit other than 0, increase the retained-digit value by 1.

(If the first digit discarded is 5, there is a simpler practice that is not uncommon but is not as effective as the preceding because of cumulative rounding errors, especially when a large number of operations is involved: If the first digit discarded is 5, increase the retained-digit value by 1.)

(c) If the first digit discarded is 6, 7, 8, or 9, increase the retained-digit value by 1.

CHAPTER 5

DETERMINING THE ERRORS OF INEXACT OR APPROXIMATE CONVERSIONS

5.1 Conversion Errors

Often, a conversion will state more than one precise value or approximate value. In such cases, the values in a given column are placed in order of decreasing accuracy.

Conversion errors may exist for inexact precise values because of uncertainty regarding the accuracy of the measurement for the last significant digit. The error becomes relatively smaller as the number of significant digits increases. If the uncertainty of the last digit of the number is ±1, the possible maximum error of the number depends on the value of that last digit in the range of 0 to 9; the greatest error is for the smallest digit, 0.

For example, any number with five significant digits is in the range of 10 000 to 99 999, regardless of decimal places; a last-digit uncertainty of ±1 would at worst cause an error of 1 part in 10 000, equal to ±0.01 %, and at best an error of ±0.001 %. For four digits, the maximum error would be ±0.1 %; for three, ±1 %; for two, ±10 %; and for one significant digit, the number 1, the maximum error would be ±100 %.

The conversion values in Tables 2 and 3 generally have five significant digits. In some instances, as for astronomy and other scientific work, a need for much greater precision may require referring to specialized sources of data.

Some precise conversions are listed with both exact and inexact conversions. In addition, the listed values that are correct only within the tolerance of +5.0 % to –5.0 %, can be useful for rough checking of detailed calculations, quick estimates, and mental calculations, and they may be relatively easy to remember. In either case, the actual magnitude of a specific error can be determined as shown in Sections 5.2 and 5.3.

5.2 Precise Conversions Having Both Exact and Inexact Values

The precise-inexact error = {the precise-inexact value minus the precise-exact value} divided by the precise-exact value.

Example: One foot = 1/3^ yard or 0.333 33 yard

$$\frac{0.333\,33 - \frac{1}{3}\hat{}}{\frac{1}{3}} \times 100 = \frac{3(0.333\,33) - \frac{3}{3}}{3(\frac{1}{3})} \times 100 = -0.001\,0\ \%\ \text{error}$$

5.3 Conversions Approximately Correct within ±5.0 %

The approximation error = {the approximation value minus the precise value} divided by the precise value.

Example: One kilometer = 0.621 37 miles, precisely, to five significant digits. Approximate conversion factors are x 5/8 or x 0.6. A 7-kilometer distance would then equal approximately 4.4 miles or 4.2 miles, respectively. The corresponding approximation errors, to the usual two significant digits, would be as follows:

$$\frac{5/8 - 0.621\ 37}{0.621\ 37} \times 100 = \frac{5 - 8(0.621\ 37)}{8(0.621\ 37)} \times 100 = +0.58\ \%\ \text{error}$$

or

$$\frac{0.6 - 0.621\ 37}{0.621\ 37} \times 100 = -3.4\ \%\ \text{error}$$

For comparison, the precise distance is 4.349 6 miles, or 4.3 miles to two significant digits.

CHAPTER 6

CHANGING THE SIZE OF SI UNITS

Metric measuring units offer great flexibility because, with prefixes, they range from extremely large to extremely small, as shown in Table 1. The flexibility can be helpful; for example: Section 3.2 derived a value of 4.832 6 E-05 millimeters per second. A more convenient but equally correct value might be 0.048 326 micrometers per second or 48.326 nanometers per second, as obtained by the following:

$$4.832\ 6\ \text{E}{-05}\ \frac{\cancel{\text{millimeters}}}{\text{second}} \times \frac{1\ 000\ \text{micrometers}}{\cancel{\text{millimeter}}} = 0.048\ 326\ \frac{\text{micrometers}}{\text{second}}$$

or

$$4.832\ 6\ \text{E}{-05}\ \frac{\cancel{\text{millimeters}}}{\text{second}} \times \frac{1\ \text{E}{+06}\ \text{nanometers}}{\cancel{\text{millimeter}}} = 48.326\ \frac{\text{nanometers}}{\text{second}}$$

Also see the discussion of the number 68 300 in Section 4.1.

CHAPTER 7

CONVERSION TABLES

Alphabetical*

** Footnotes in random order due to alphabetical sort. They appear in numerical order in "Group" Table 3.*

TABLE 2
Measurement Conversions, Alphabetical

All measurement units are US, unless otherwise noted. All number denominations "billion" and higher are US, unless otherwise noted.

CONVERT	TO EQUIVALENT	BY PRECISELY	OR WITHIN ± 5.0 %
-A-			
abampere	amperes	x 10^	
	coulombs per second	x 0.1^	
	faradays (based on carbon-12) per second	x 1.036 4 E-04	*x 1 E-04*
	faradays (chemical) per second	x 1.036 3 E-04	*x 1 E-04*
	faradays (physical) per second	x 1.036 0 E-04	*x 1 E-04*
	statamperes	x 2.997 9 E+10	*x 3 E+10*
abampere-turn	ampere-turns	x 10^	
	gilberts	x 4 pi or x 12.566	*x 1.3*
abampere-turn per centimeter	ampere-turns per centimeter	x 10^	
	ampere-turns per inch	x 25.4^	*x 25*
	oersteds	x 4 pi or x 12.566	*x 13*
abcoulomb	ampere-hours	/360^ or x 0.002 777 8	*x 0.002 8*
	coulombs	x 10^	
	faradays (based on carbon-12)	x 1.036 4 E-04	*x 1 E-04*
	statcoulombs	x 2.997 9 E+10	*x 3 E+10*
abfarad	farads	x 1^E+09	
abhenry	henrys	x 1^E-09	
abmho	mhos	x 1^E+09	
	siemens	x 1^E+09	
abohm	ohms	x 1^E-09	
abohm per centimeter cube	abohm-centimeter	x 1^	
absolute alcohol	ethanol-water solution containing 5.1 percent , or less, water by volume	x 1^	
absolute value of a number	magnitude of the number regardless of algebraic sign	x 1^	
absolute value of vector	magnitude regardless of direction	x 1^	
absolute zero temperature	deg. Rankine	-273.15	
	kelvins	x 0^	
absorbed dose	gray	x 1^	
absorbed dose rate	gray per second	x 1^	
abvolt	volts	x 1^E-08	
acre	ares	x 40.469	*x 40*
	forties	/40^ or x 0.025^	
	hectares	x 0.404 69	*x 0.4*
	quarter sections	/160^ or x 0.006 25^	*x 0.006*
	square chains, surveyor's	x 10^	
	square feet	x 43 560^	*x 43 000*
	square kilometers	x 0.004 046 9	*x 0.004*
	square links, surveyor's	x 1^E+05	
	square meters	x 4 046.9	*x 4 000*
	square miles	/640^ or x 0.001 562 5^	*x 0.001 6*
	square rods	x 160^	
	square yards	x 4 840^	*x 5 000*
acre (Can.)	acre (US)	x 1^	
acre-foot	cubic feet	x 43 560^	*x 44 000*
	cubic meters	x 1 233.5	*x 1 200*
	cubic yards	x 1 613.3	*x 1 600*
	gallons	x 3.258 5 E+05	*x 3.3 E+05*
acre-foot per day	acre-inches per hour	/2^ or x 0.5^	
	gallons per minute	x 226.29	*x 220*

TABLE 2
Measurement Conversions, Alphabetical

All measurement units are US, unless otherwise noted. All number denominations "billion" and higher are US, unless otherwise noted.

CONVERT	TO EQUIVALENT	BY PRECISELY	OR WITHIN ± 5.0 %
acre-foot per hour	cubic feet per hour	x 43 560^	*x 44 000*
	gallons per minute	x 5 430.9	*x 5 400*
acute angle	angle less than 90 degrees	x 1	
admittance	1/impedance	x 1^	
agate line (in newspaper advertisements)	one column wide, 1/14-inch deep	x 1	
agate (printer's type)	points (printer's)	x 5.5	
ambient pressure	environmental pressure [11]	x 1^	
ampere	abamperes	/10^ or x 0.1^	
	coulomb per second	x 1^	
	EMU of current	/10^ or x 0.1^	
	ESU of current	x 2.997 9 E+09	*x 3 E+09*
	gilberts	x 1.256 6	*x 1.3*
	statamperes	x 2.997 9 E+09	*x 3 E+09*
ampere per meter	oersteds	x 0.012 566	*x 0.013*
ampere per square centimeter	amperes per square inch	x 6.451 6^	*x 6.5*
	amperes per square meter	x 1^E+04	
ampere per square inch	amperes per square centimeter	x 0.155 00	*x 0.16*
	amperes per square meter	x 1 550.0	*x 1 500*
ampere per square meter	amperes per square centimeter	x 1^E-04	
	amperes per square inch	x 6.451 6^E-04	*x 6.5 E-04*
ampere, absolute	amperes, international	x 1.000 2	*x 1*
ampere-hour	abcoulombs	x 360^	
	coulombs	x 3 600^	
	faradays (based on carbon-12)	x 0.037 311	*x 0.037*
	faradays, chemical	x 0.037 307	*x 0.037*
	faradays, physical	x 0.037 297	*x 0.037*
	statcoulombs	x 1.079 3 E+13	*x 1.1 E+13*
ampere-turn	abampere-turns	/10^ or x 0.1^	
	gilberts	x 0.4 pi or x 1.256 6	*x 1.3*
ampere-turn per centimeter	abampere-turns per centimeter	/10^ or x 0.1^	
	ampere-turns per inch	x 2.54^	*x 2.5*
	oersted	x 0.4 pi or x 1.256 6	*x 1.3*
ampere-turn per inch	abampere-turns per centimeter	x 5/127^ or x 0.0393 70	*x 0.04*
	ampere-turns per centimeter	x 50/127^ or x 0.393 70	*x 0.4*
	ampere-turns per meter	x 5 000/127^ or x 39.370	*x 40*
	gilberts per centimeter	x 0.494 74	*x 0.5*
ampere-turn per meter	ampere-turns per centimeter	x 0.01^	
	ampere-turns per inch	x 0.025 4^	*x 0.025*
	gilberts per centimeter	x 0.012 566	*x 0.013*
	oersteds	x 0.012 566	*x 0.013*
amu	atomic mass unit	x 1^	
angstrom	centimeters	x 1^E-08	
	inches	x 3.937 0 E-09	*x 4 E-09*
	meters	x 1^E-10	
	microns	x 1^E-04	
API scale (for petroleum)	specific gravity, in deg. API	x 1^	

[11] Ambient pressure is the environmental pressure surrounding a device and is not necessarily atmospheric pressure, whether standard or local.

TABLE 2
Measurement Conversions, Alphabetical

All measurement units are US, unless otherwise noted. All number denominations "billion" and higher are US, unless otherwise noted.

CONVERT	TO EQUIVALENT	BY PRECISELY	OR WITHIN ± 5.0 %
are	acres	x 0.024 710	*x 0.025*
	square feet	x 1 076.4	*x 1 100*
	square meters	x 100^	
	square yards	x 119.60	*x 120*
arpent (French land area, Queb.)	square feet (French land measure, Queb.)	x 32 400^	*x 32 000*
arpent (French land length, Queb.)	feet (French land measure, Queb.)	x 180^	
astronomical unit	kilometers	x 1.496 0 E+08	*x 1.5 E+08*
	light-year	x 1.581 3 E-05	*x 1.6 E-05*
	meters	x 1.496 0 E+11	*x 1.5 E+11*
	parsec	x 4.848 1 E-06	*x 5 E-06*
atmosphere, standard[12]	bars	x 1.013 3	*x 1*
	centimeters of mercury [13]	x 76.000	*x 76*
	centimeters of water [13]	x 1 033.3	*x 1 000*
	feet of water [13]	x 33.900	*x 34*
	inches of mercury [13]	x 29.921	*x 30*
	kilograms (force) per square centimeter	x 1.033 2	*x 1*
	kilograms (force) per square meter	x 10 332	*x 10 000*
	kilopascals	x 101.33	*x 100*
	meters of mercury [13]	x 0.760 00	*x 0.76*
	millibars	x 1 013.3	*x 1 000*
	millimeters of mercury [13]	x 760.00	*x 760*
	newtons per square meter	x 1.013 25 E+05	*x 1 E+05*
	pounds (force) per square inch	x 14.696	*x 14.7 or x 15*
	tons, short (force) per square foot	x 1.058 1	*x 1.1*
	tons, short (force) per square inch	x 0.007 348 0	*x 0.007*
	torrs	x 760.00	*x 760*
atmosphere, technical	kilograms (force) per square centimeter	x 1^	
	pascals	x 9.806 7 E+04	*x 10 E+04*
atmospheric pressure	the omnidirectional pressure created at any specific location by the weight of the atmosphere	x 1^	
atomic mass unit	electron-volts (equivalent energy)	x 9.314 8 E+08	*x 9 E+08*
	ergs (equivalent energy)	x 0.001 492 4	*x 0.001 5*
	grams	x 1.660 5 E-24	*x 1.7 E-24*
	mass of carbon-12 atom	/12^ or x 0.083 333	*x 0.08*
atomic mass, relative	atomic weight	x 1^	
audible sound	See "range"		
aught	zero	x 1^	
Avogadro constant	Avogadro number	x 1^	
Avogadro number	6.022 14 E+23	x 1	*6 E+23*
	number of atoms per gram-atom	x 6.022 14 E+23	*x 6 E+23*
	number of molecules per gram-molecule	x 6.022 14 E+23	*x 6 E+23*
avogram	grams	1/Avogadro number or x 1.660 54 E-24	*x 1.7 E-24*
-B-			
baby (for wine, Brit.)	bottles (Brit.)	/8^ or x 0.125^	*x 0.13*
	liters	x 3/32 or x 0.093 75	*x 0.09*

[12] Standard atmospheric conditions are: acceleration of gravity = 9.806 65 meters per second per second = 32.174 0 feet per second per second; atmospheric pressure = 760.00 centimeters of mercury = 29.921 3 inches of mercury = 14.695 9 pounds (force) per square inch; temperature = 0.0 deg. C = 32.0 deg. F.

[13] Unless otherwise noted, liquid-head conversions are based on: a pressure of one standard atmosphere; temperature for mercury = 0.0 deg. C = 32.0 deg. F; temperature for water = 4.0 deg. C = 39.2 deg. F.

TABLE 2
Measurement Conversions, Alphabetical

All measurement units are US, unless otherwise noted. All number denominations "billion" and higher are US, unless otherwise noted.

CONVERT	TO EQUIVALENT	BY PRECISELY	OR WITHIN ± 5.0 %
bag (of cement)	pounds, net	x 94^	*x 90*
bale (of cotton)	pounds	x 500^	
Balling scale (for brewing industry)	dissolved-solids weight percentages, in deg. Balling	x 1^	
balthazar (for wine, Brit.)	bottles (Brit.)	x 16^	
	liters	x 12	
bar	atmospheres, standard	x 0.986 92	*x 1*
	dynes per square centimeter	x 1^E+06	
	kilograms (force) per square centimeter	x 1.019 7	*x 1*
	kilopascals	x 100^	
	newtons per square meter	x 1^E+05	
	pascals	x 1^E+05	
	pounds (force) per square foot	x 2 088.5	*x 2 000*
	pounds (force) per square inch	x 14.504	*x 15*
barn	square meters	x 1^E-28	
baron (of meat)	sirloins or loins	x 2^	
barrel (42 gallons) per day	barrels (42 gallons) per hour	/24^ or x 0.041 667	*x 0.04*
	cubic feet per hour	x 0.233 94	*x 0.23*
	cubic feet per minute	x 0.003 899 0	*x 0.004*
	cubic feet per second	x 6.498 4 E-05	*x 6.5 E-05*
	gallons per hour	x 1.750 0	*x 1.8*
	gallons per minute	x 7/240^ or x 0.029 167	*x 0.03*
	liters per second	x 0.001 840 1	*x 0.001 8*
barrel (42 gallons) per hour	barrels (42 gallons) per day	x 24^	
	cubic feet per hour	x 5.614 6	*x 5.6*
	cubic feet per minute	x 0.093 576	*x 0.09*
	cubic feet per second	x 0.001 559 6	*x 0.001 5*
	gallons per hour	x 42^	
	gallons per minute	x 0.7^	
	liters per second	x 0.044 163	*x 0.044*
barrel (Can.)	gallon (Can.)	x 36^	
barrel (for cement)	pounds, net	x 376^	*x 380*
barrel (for cisterns, in one State)	gallons	x 36^	
barrel (for cranberries)	bushels, struck measure	x 2.709	*x 2.7*
	cubic inches	x 5 826	*x 6 000*
	quarts, dry	x 5 549/64^ or x 86.703	*x 90*
barrel (for fermented liquor, Federal)	gallons	x 31^	*x 30*
barrel (for flour)	pounds	x 196^	*x 200*
barrel (for fruits, vegetables, except cranberries)	bushels, struck measure	x 3.281	*x 3.3*
	cubic inches	x 7 056	*x 7 000*
	quarts, dry	x 105	*x 100*
barrel (for lime), standard large	pounds, net	x 280^	
barrel (for lime), standard small	pounds, net	x 180^	
barrel (for liquids, in four States)	gallons	x 42^	
barrel (for liquids, in many States)	gallons	x 31.5^	*x 32*
barrel (for petroleum)	gallons	x 42^	

TABLE 2
Measurement Conversions, Alphabetical

All measurement units are US, unless otherwise noted. All number denominations "billion" and higher are US, unless otherwise noted.

CONVERT	TO EQUIVALENT	BY PRECISELY	OR WITHIN ± 5.0 %
barrel (for petroleum) (42 gallons)	cubic meters	x 0.158 99	*x 0.16*
barrel (for proof spirits, Federal)	gallons	x 40^	
barrel (for sand, Can.)	liters	x 81.830	*x 80*
barrel (per service and law)	gallons	x 31 to x 42	
barrelage	number of barrels	x 1	
barrel, herring (Can.)	liters	x 145.47	
barrel, Imperial (Brit.)	bushels, Imperial (Brit.)	x 4.5^	
	kilderkins (Brit.)	x 2	
barrel, Imperial (for ships' water, Brit.)	gallons, Imperial (Brit.)	x 36^	
barye	dyne per square centimeter	x 1^	
base number	See "radix"		
base point	See "radix point"		
basis point (for loan investment yields)	0.01^ of one percent	x 1^	
baud	bit per second	x 1^	
	characters per second	/8^ or x 0.125^	*x 0.13*
Baume scales (for light and heavy liquids)	specific gravities, in deg. Baume	x 1^	
beat frequency	frequency difference of two slightly different simultaneous tones	x 1^	
Beaufort scale (for wind)	See APPENDIX, "wind speeds"		
becquerel	curies	x 2.702 7 E-11	*x 2.7 E-11*
	radionuclide activity	x 1/second	
bel	decibels	x 10^	
	log of ratio of two power levels	x 1	
ber	bit error rate	x 1^	
bicentennial	years	x 200^	
biennium	years	x 2^	
bilateral	sides	x 2^	
billion (Brit.)	trillion (US)	x 1^	
billion (US)	1^E+09	x 1^	
billisecond	nanosecond	x 1^	
binary	based on the number 2	x 1^	
biot	amperes	x 10^	
bipartite	parts	x 2^	
bips	billion (US) instructions per second	x 1^	
bissextile	a leap year	x 1^	
bit	binary digit	x 1^	
	bytes	/8^ or x 0.125^	*x 0.13*
	nibbles	/4^ or x 0.25^	
board foot	cubic feet (nominal)	/12^ or x 0.083 333	*x 0.08*
	cubic inches (actual)	x 96 (approx.)	*x 100*
	cubic inches (nominal)	x 144	
	liters (nominal)	x 2.359 7	*x 2 400*
Bohr magneton	joule-square meters per weber	x 9.28 E-24	*x 9 E-24*
boiling point of water (ITS)	deg. Celsius (@ 101-325 kPa)	x 100^	
	deg. Fahrenheit (@ 14.696 psia)	x 212^	*x 210*
bolt (for cloth) (US)	meters	x 18.3 to x 36.6 (usually)	
	yards	x 20 to 40 (usually)	

TABLE 2
Measurement Conversions, Alphabetical

All measurement units are US, unless otherwise noted. All number denominations "billion" and higher are US, unless otherwise noted.

CONVERT	TO EQUIVALENT	BY PRECISELY	OR WITHIN ± 5.0 %
bone dry	percent moisture	zero (nominal)	
bottle (for wine, Brit.)	liters	x 3/4^ or x 0.75^	
bpd	barrel per day	x 1^	
bph	barrel per hour	x 1^	
bpi	bits per inch	x 1^	
bpm	barrel per minute	x 1^	
bps	bits per second	x 1^	
brace	similar things	x 2^	
British thermal unit	Btu	x 1^	
Brix scale (for sugar industry)	sucrose weight percentages in water solution, in deg. Brix	x 1^	
Btu	British thermal unit	x 1^	
	centigrade heat units	x 5/9 or x 0.555 56	*x 0.56*
Btu per hour	tons of refrigeration	/12 000 or x 8.333 3 E-05	*x 8 E-05*
Btu per minute	tons of refrigeration	/200^ or x 0.005^	
Btu (IT)	calories (IT)	x 252.00	*x 250*
	ergs	x 1.055 1 E+10	*x 1.1 E+10*
	foot-pounds (force)	x 778.17	*x 800*
	horsepower-hours	x 3.930 1 E-04	*x 4 E-04*
	joules	x 1 055.1	*x 1 100*
	kilocalories (IT)	x 0.252 00	*x 0.25*
	kilowatt-hours	x 2.930 7 E-04	*x 3 E-04*
	liter-atmospheres, standard	x 10.413	*x 10*
	meter-kilograms (force)	x 107.59	*x 110*
Btu (IT) per day-square foot	Btu (IT) per hour-square foot	/24^ or x 0.041 667	*x 0.04*
	calories (IT) per hour-square centimeter	x 0.011 302	*x 0.011*
	calories (IT) per second-square centimeter	x 3.139 4 E-06	*x 3 E-06*
	watts per square centimeter	x 1.314 4 E-05	*x 1.3 E-05*
Btu (IT) per day-square foot-deg. F	Btu (IT) per hour-square foot-deg. F	/24^ or x 0.041 667	*x 0.04*
	calories (IT) per hour-square centimeter-deg. C	x 0.020 343	*x 0.02*
	calories (IT) per second-square centimeter-deg. C	x 5.651 0 E-06	*x 5.7 E-06*
	watts per square centimeter-deg. C	x 2.365 9 E-05	*x 2.4 E-05*
Btu (IT) per day-square foot-deg. F/inch	Btu (IT) per hour-square foot-deg. F/foot	/288^ or x 0.003 472 2	*x 0.003 5*
	calories (IT) per hour-square centimeter-deg. C/centimeter	x 0.051 672	*x 0.05*
	calories (IT) per second-square centimeter-deg. C/centimeter	x 1.435 4 E-05	*x 1.4 E-05*
	watts per square centimeter-deg. C/centimeter	x 6.009 5 E-05	*x 6 E-05*
Btu (IT) per hour	calories (IT) per second	x 4.199 9	*x 4*
	foot-pounds (force) per second	x 0.216 16	*x 0.22*
	horsepower	x 3.930 1 E-04	*x 4 E-04*
	watts	x 0.293 07	*x 0.3*
Btu (IT) per hour-square foot	Btu (IT) per day-square foot	x 24^	
	calories (IT) per hour-square centimeter	x 0.271 25	*x 0.27*
	calories (IT) per second-square centimeter	x 7.534 6 E-05	*x 7.5 E-05*
	watts per square centimeter	x 3.154 6 E-04	*x 3.2 E-04*
	watts per square meter	x 3.154 6	*x 3.2*

TABLE 2
Measurement Conversions, Alphabetical

All measurement units are US, unless otherwise noted. All number denominations "billion" and higher are US, unless otherwise noted.

CONVERT	TO EQUIVALENT	BY PRECISELY	OR WITHIN ± 5.0 %
Btu (IT) per hour-square foot-deg. F	Btu (IT) per day-square foot-deg. F	x 24^	
	calories (IT) per hour-square centimeter-deg. C	x 0.488 24	*x 0.5*
	calories (IT) per second-square centimeter-deg. C	x 1.356 2 E-04	*x 1.4 E-04*
	watts per square centimeter-deg. C	x 5.678 3 E-04	*x 5.7 E-04*
Btu (IT) per hour-square foot-deg. F/foot	Btu (IT) per day-square foot-deg. F/inch	x 288^	*x 300*
	calories (IT) per hour-square centimeter-deg. C/centimeter	x 14.882	*x 15*
	calories (IT) per second-square centimeter-deg. C/centimeter	x 0.004 133 8	*x 0.004*
	watts per square centimeter-deg. C/centimeter	x 0.017 317	*x 0.017*
Btu (IT) per kilowatt-hour (power-plant heat rate)	joules per kilowatt-hour	x 1 055.1	*x 1 100*
Btu (IT) per minute	foot-pounds (force) per second	x 12.969	*x 13*
	horsepower	x 0.023 581	*x 0.024*
	kilowatts	x 0.0175 84	*x 0.018*
	watts	x 17.584	*x 18*
Btu (IT) per minute-square foot	watts per square inch	x 0.122 11	*x 0.12*
Btu (IT) per second	watts	x 1 055.1	*x 1 100*
Btu (IT) per second-square foot	watts per square meter	x 11 357	*x 11 000*
Btu (thermochemical)	joules	x 1 054.4	*x 1 100*
Btu (thermochemical) per hour	watts	x 0.292 88	*x 0.3*
Btu (thermochemical) per hour per square foot	watts per square meter	x 3.152 5	*x 3.2*
Btu (thermochemical) per minute	watts	x 17.573	*x 18*
Btu (thermochemical) per minute-square foot	watts per square meter	x 189.15	*x 190*
Btu (thermochemical) per second	watts	x 1 054.4	*x 1 100*
Btu (thermochemical) per second-square foot	watts per square meter	x 11.349	*x 11*
Btu (thermochemical) per second-square inch	watts per square meter	x 1.634 2 E+06	*x 1.6 E+06*
Btu (@ 39 deg. F)	joules	x 1 059.7	*x 1 100*
Btu (@ 59 deg. F)	joules	x 1 054.8	*x 1 100*
Btu (@ 60 deg. F)	joules	x 1 054.7	*x 1 100*
Btu (@ 60.5 deg. F, Can.) [4]	joules	x 1 054.6	*x 1 100*
Btu, mean (for range of 0 to 100 deg.C)	joules	x 1 055.9	*x 1 100*
bundle (builder's)	square feet	x 20^, x 25^, or x 100/3^ (per type of shingle)	
	square (builder's)	/3^ or x 0.333 33	*x 0.33*
bundle (for papermaking)	pounds	x 50	
bundle (for shipping paper)	pounds	x 125	*x 130*

[4] The Btu (@ 60.5 deg. F) is used by the Canadian petroleum and natural-gas industry.

TABLE 2
Measurement Conversions, Alphabetical

All measurement units are US, unless otherwise noted. All number denominations "billion" and higher are US, unless otherwise noted.

CONVERT	TO EQUIVALENT	BY PRECISELY	OR WITHIN ± 5.0 %
bushel (Can.)	bushel, Imperial (Brit.)	x 1^	
bushel (customary)	bushel, struck measure	x 1^	
bushelage	number of bushels	x 1	
bushel, heaped	bushels, struck measure	x 1.277 8 (often x 1 1/4)	*x 1.3*
	cubic feet	x 1.590 1	*x 1.6*
	cubic inches	x 2 747.7	*x 2 700*
bushel, Imperial struck measure (Brit.)	bushels, struck measure (US)	x 1.032 1	*x 1*
	cubic feet	x 1.284 4	*x 1.3*
	cubic inches	x 2 219.4	*x 2.2*
bushel, Imperial (Brit.)	gallons, Imperial (Brit.)	x 8^	
	pecks (Brit.)	x 4^	
bushel, struck measure	bushels, heaped	x 0.782 62	*x 0.8*
	cubic feet	x 1.244 5	*x 1.2*
	cubic inches	x 2 150.4	*x 2 200*
	cubic meters	x 0.035 239	*x 0.035*
	liters	x 35.239	*x 35*
bushel, Winchester struck (Brit.)	cubic inches	x 2 150.4	*x 2 200*
but	See "butt"		
butt	gallons	x 126	*x 130*
	hogsheads	x 2^	
	liters	x 476.96	*x 500*
butt (Brit.)	gallons, Imperial (Brit.)	x 108	*x 100*
	hogsheads (Brit.)	x 2^	
butt (for ship's water, Brit.)	gallons, Imperial (Brit.)	x 110^	
byte	bits	x 8^	
	nibbles	/2^ or x 0.5^	
-C-			
cable	See "cable length"		
cable length	fathoms	x 120^	
	feet	x 720^	
	meters	x 219.46	*x 220*
	miles	x 3/22^ or x 0.136 36	*x 0.14*
cable length (Brit.)	feet	x 608^	*x 600*
	miles	x 0.115 15	*x 0.12*
calendar, Gregorian (since 1582) [19]	calendar (customary)	x 1^	
calendar, Julian day	See "Julian day calendar"		
calendar, Julian (superseded 1582)	calendar, Gregorian (since 1582)	+10 days initially [22]	
calendar, New Style	calendar, Gregorian (since 1582)	x 1^	
calendar, Old Style	calendar, Julian (superseded 1582)	x 1^	
calorie	kilocalories	/1 000^ or x 0.001^	
Calorie (capitalized)	kilocalories (usually, also see "calorie" (not capitalized)	x 1	

[19] In the Gregorian calendar, every year whose number is divisible by four is a leap year except for centesimal years that are not exactly divisible by 400. The year 2000 is a leap year, 2100 is not a leap year.

[22] Gregorian calendar dates minus Julian calendar dates equals + 10 days in the period 1582 to 1700; + 11 days, 1700 to 1800; + 12 days, 1800 to 1900; + 13 days, 1900 to 2100. George Washington's birthday is February 22, 1742 (Gregorian), equivalent to February 11, 1732 (Julian).

TABLE 2
Measurement Conversions, Alphabetical

All measurement units are US, unless otherwise noted. All number denominations "billion" and higher are US, unless otherwise noted.

CONVERT	TO EQUIVALENT	BY PRECISELY	OR WITHIN ± 5.0 %
calorie [5]	See "kilocalorie"		
calorie (IT)	Btu (IT)	x 0.003 968 3	*x 0.004*
	calories, thermochemical	x 1.000 7	*x 1*
	foot-pounds (force)	x 3.088 0	*x 3*
	joules	x 4.186 8^	*x 4*
	meter-kilograms (force)	x 0.426 93	*x 0.43*
calorie (IT) per hour-square centimeter	Btu (IT) per day-square foot	x 88.481	*x 90*
	Btu (IT) per hour-square foot	x 3.686 7	*x 3.7*
	calories (IT) per second-square centimeter	/3 600^ or x 2.777 8 E-04	*x 2.8 E-04*
	watts per square centimeter	x 0.001 163 0	*x 0.001 2*
calorie (IT) per hour-square centimeter-deg. C	Btu (IT) per day-square foot-deg. F	x 49.156	*x 50*
	Btu (IT) per hour-square foot-deg. F	x 2.048 2	*x 2*
	calories (IT) per second-square centimeter-deg. C	/3 600^ or x 2.777 8 E-04	*x 2.8 E-04*
	watts per square centimeter-deg. C	x 0.001 163 0	*x 0.001 2*
calorie (IT) per hour-square centimeter-deg. C/centimeter	Btu (IT) per day-square foot-deg. F/inch	x 19.353	*x 20*
	Btu (IT) per hour-square foot-deg. F/foot	x 0.067 197	*x 0.07*
	calories (IT) per second-square centimeter-deg. C/centimeter	/3 600^ or x 2.777 8 E-04	*x 2.8 E-04*
	watts per square centimeter-deg. C/centimeter	x 0.001 163 0	*x 0.001 2*
calorie (IT) per second-square centimeter	Btu (IT) per day-square foot	x 3.185 3 E+05	*x 3.2 E+05*
	Btu (IT) per hour-square foot	x 13 272	*x 13 000*
	calories (IT) per hour-square centimeter	x 3 600^	
	watts per square centimeter	x 4.186 8^	*x 4*
calorie (IT) per second-square centimeter-deg. C	Btu (IT) per day-square foot-deg. F	x 1.769 6 E+05	*x 1.8 E+05*
	Btu (IT) per hour-square foot-deg. F	x 7 373.4	*x 7 400*
	calories (IT) per hour-square centimeter-deg. C	x 3 600^	
	watts per square centimeter-deg. C	x 4.186 8^	*x 4*
calorie (IT) per second-square centimeter-deg. C/centimeter	Btu (IT) per day-square foot-deg. F/inch	x 69 670	*x 70 000*
	Btu (IT) per hour-square foot-deg. F/foot	x 241.91	*x 240*
	calories (IT) per hour-square centimeter-deg. C/centimeter	x 3 600^	
	watts per square centimeter-deg. C/centimeter	x 4.186 8^	*x 4*
calorie (not capitalized)	calories (usually, also see "Calorie (capitalized)")	x 1^	
calorie (thermochemical) per minute-square centimeter	watts per square meter	x 697.33	*x 700*
calorie (thermochemical) per second	watts	x 4.184^	*x 4*

[5] The energy value of food and drink is customarily stated in "calories", but the technically correct measuring unit is "kilocalories", which is sometimes called "large calories". (1 kilcalorie = 1 000 calories) A person on a reducing diet may take 1 100 "calories" per day, but, in truth, he is taking 1 100 kilocalories, or 1.1 million (1 100 000) calories, or 1.1 megacalories.

TABLE 2
Measurement Conversions, Alphabetical

All measurement units are US, unless otherwise noted. All number denominations "billion" and higher are US, unless otherwise noted.

CONVERT	TO EQUIVALENT	BY PRECISELY	OR WITHIN ± 5.0 %
calorie (@ 15 deg. C)	joules	x 4.185 8	*x 4*
calorie (@ 20 deg. C)	joules	x 4.181 9	*x 4*
calorie, gram	calorie	x 1^	
calorie, kilogram	kilocalorie	x 1^	
calorie, large	kilocalorie	x 1^	
calorie, mean (for range 0 to 100 deg. C)	joules	x 4.190 0	*x 4*
calorie, Ostwald	kilocalorie	/10^ or x 0.1^	
calorie, small	calorie, gram	x 1^	
calorie, thermochemical	calories (IT)	x 0.999 33	*x 1*
	joules	x 4.184^	*x 4*
calory	calorie	x 1^	
candela per square centimeter	lamberts	x pi or x 3.141 6	*x 22/7 or x 3*
candela per square inch	candelas per square meter	x 1 550.0	*x 1 600*
candela per square meter	candelas per square inch	x 6.451 6 E-04	*x 6.5 E-04*
	footlamberts	x 0.29 186	*x 0.3*
	stilbs	/10 000^ or x 1^E-04	
candle	candelas	x 1.02	*x 1*
candle per square centimeter	candles per square inch	x 6.451 6^	*x 6.5*
candlepower	See "spherical candlepower"		
candlepower, spherical	See "spherical candlepower"		
candle, hefner	candela	x 0.92	*x 0.9*
candle, international	candle	x 1^	
	lumen per steradian	x 1^	
candle, new	candela	x 1^	
	candlepower per square centimeter	x 60^	
candle, standard	candela	x 1^	
capacity, deadweight	tonnage, deadweight	x 1^	
carat grain (for pearls)	carats	/4^ or x 0.25^	
carat (for gemstones)	grains	x 3.086 5	*x 3*
	ounces, apothecary	x 0.006 430 1	*x 0.006 4*
	ounces, avoirdupois	x 0.007 054 8	*x 0.007*
	ounces, troy	x 0.006 430 1	*x 0.006 4*
	points (jeweler's)	x 100^	
carat (for gold)	See "karat"		
carat (international)	milligrams	x 200^	
carat (US before 1913)	milligrams	x 205.3	*x 200*
carat (US since 1913)	milligrams	x 200^	
carat, metric (for precious metals)	carat (intl.)	x 1^	
	grains	x 3.086 5	*x 3*
carboy	gallons	x 5 to x 15	
cart, salt (Can.)	liters	x 490.98	*x 500*
cart, tub (Can.)	liters	x 81.830	*x 80*
carucate (Eng.)	hide	x 1	
cent per cent	a hundred for each hundred	x 1^	
cent (sound)	interval between two frequencies having ratio of 1 200th root of 2 (= 1.0006)	x 1^	
centage	percentage	x 1^	

TABLE 2
Measurement Conversions, Alphabetical

All measurement units are US, unless otherwise noted. All number denominations "billion" and higher are US, unless otherwise noted.

CONVERT	TO EQUIVALENT	BY PRECISELY	OR WITHIN ± 5.0 %
cental (Brit.)	hundredweights, short (US)	x 1^	
	pounds, avoirdupois	x 100^	
cental (Can.)	cental (Brit.)	x 1^	
centare	square meter	x 1^	
centenary	years	x 100^	
centennial	years	x 100^	
centesimal	divided into hundredths	x 1^	
centiare	centare	x 1^	
centigrade heat unit	Btu	x 1.8	
	calories (IT)	x 453.59	*x 450*
	Chu	x 1^	
	joules	x 1 899.1	*x 1 900*
	meter-kilograms (force)	x 193.65	*x 200*
centillion (Brit.)	1^E+600	x 1^	
centillion (US)	1^E+303	x 1^	
centimeter	angstroms	x 1^E+08	
	feet	x 0.032 808	*x 0.032*
	inches	/2.54^ or x 0.393 70	*x 0.4*
	meters	/100^ or x 0.01^	
	miles	x 6.213 7 E-06	*x 6 E-06*
	mils	x 393.70	*x 400*
	yards	x 0.010 936	*x 0.011*
centimeter of mercury [13]	pascals	x 1 333.2	*x 1 300*
	pounds (force) per square foot	x 27.845	*x 28*
	pounds (force) per square inch	x 0.193 37	*x 0.19*
centimeter of water [13]	pascals	x 98.064	*x 100*
centimeter per second per second	inches per second per second	x 0.393 70	*x 0.4*
centipoise	pascal-seconds	/1 000^ or x 0.001^	
	pound (force)-seconds per square foot	x 2.088 5 E-05	*x 2.1 E-05*
	pounds (mass) per foot-second	x 6.719 7 E-04	*x 7 E-04*
centistoke	square feet per second	x 1.076 4 E-05	*x 1.1 E-05*
	square meters per second	x 1^E-06	
	square millimeters per second	x 1^	
centner	kilograms	x 50^	
centner (for assaying)	dram	x 1^	
centner, double	centner, metric	x 1^	
centner, metric	kilograms	x 100^	
centuple	100 times as large	x 1^	
century	years	x 100^	
cetane number (diesel-fuel ignition rating)	volume percent of cetane in a standard reference fuel	x 1^	
cetane rating	cetane number	x 1^	
cfd	cubic foot per day	x 1^	
cfh	cubic foot per hour	x 1^	
cfm	cubic foot per minute	x 1^	
cfs	cubic foot per second	x 1 ^	
chain (Can.)	chain, surveyor's (US)	x 1^	
chain (for football, US)	yards	x 10^	

[13] Unless otherwise noted, liquid-head conversions are based on: a pressure of one standard atmosphere; temperature for mercury = 0.0 deg. C = 32.0 deg. F; temperature for water = 4.0 deg. C = 39.2 deg. F.

TABLE 2
Measurement Conversions, Alphabetical

All measurement units are US, unless otherwise noted. All number denominations "billion" and higher are US, unless otherwise noted.

CONVERT	TO EQUIVALENT	BY PRECISELY	OR WITHIN ± 5.0 %
chain, engineer's	feet	x 100^	
	links	x 100^	
chain, Gunter's	chain, surveyor's	x 1^	
chain, Ramden's	chain, engineer's	x 1^	
chain, surveyor's	feet	x 66^	
	links	x 100^	
	meters	x 20.117	*x 20*
	miles	/80^ or x 0.012 5^	*x 0.013*
	rods (length)	x 4^	
chaldron	bushels	x 36	
	cubic meters	x 1.268 6	*x 1.3*
chaldron (Eng.)	bushels, Imperial (Eng.)	x 32 to x 72	
chaldron (for coal, Eng.)	bushels, Imperial (Eng.)	x 36	
	hundredweights (Eng.)	x 25.5	*x 26*
Chandler wobble (of Earth around axis)	days	x 440 (approx.)	
chiliad (quantity)	items	x 1 000^	
chiliad (time)	years	x 1 000^	
Chu	centigrade heat unit	x 1^	
cipher	zero	x 1^	
circle	degrees (angle)	x 360^	
	points (compass)	x 32^	
circular inch	circular mils	x 1^E+06	
	square inches	x pi/4 or x 0.785 40	*x 0.8*
circular mil	circular inches	x 1^E-06	
	square feet	x 5.454 2 E-09	*x 5.5 E-09*
	square inches	x 7.854 0 E-07	*x 8 E-07*
	square meters	x 5.067 1 E-10	*x 5 E-10*
	square mils	x pi/4 or x 0.785 40	*x 0.8*
	square yards	x 6.060 2 E-10	*x 6 E-10*
cistern (4 x 2.5 x 3 feet, Brit.)	gallons, Imperial (Brit.)	x 186.96	*x 190*
clove (Eng.)	pounds	x 8	
clusec (vacuum-pumping power)	watts	x 1.333 E-06	*x 1.3 E-06*
coefficient of heat transfer	thermal conductance	x 1^	
colloidal-particle size range	meters	x 1^E-09 to 1^E-06	
color temperature	See APPENDIX, "temperature color scale"		
concentration, molal	moles of solute per kilogram of solvent	x 1^	
concentration, molar	moles of solute per liter of solution	x 1^	
concentration, normal	gram-equivalent of solute per liter of solution	x 1^	
concentration, percent (by mass)	grams of solute per 100 grams of solution	x 1^	
concentration, percent (by volume)	milliliters of solute per 100 milliliters of solution	x 1^	
concentration, relative	moles per cubic meter	x 1^	
confidence level (statistical)	See APPENDIX, "numbers"		
coomb (Eng.)	bushels, Imperial (Eng.)	x 4	
cord foot (for stacked wood, 4 x 4 x 1 feet)	cords	/8 or x 0.125	*x 0.13*

TABLE 2
Measurement Conversions, Alphabetical

All measurement units are US, unless otherwise noted. All number denominations "billion" and higher are US, unless otherwise noted.

CONVERT	TO EQUIVALENT	BY PRECISELY	OR WITHIN ± 5.0 %
	cubic feet	x 16	
cord (for stacked wood, 4 x 4 x 8 feet)	cubic feet	x 128	*x 130*
	cubic meters	x 3.624 6	*x 3.6*
coulomb	abcoulombs	/10^ or x 0.1^	
	ampere-hours	/3 600^ or x 2.777 8 E-04	*x 2.8 E-04*
	ampere-second	x 1^	
	faradays (based on carbon-12)	x 1.036 4 E-05	*x 1 E-05*
	faradays, chemical	x 1.036 3 E-05	*x 1 E-05*
	faradays, physical	x 1.036 0 E-05	*x 1 E-05*
	statcoulombs	x 2.997 9 E+09	*x 3 E+09*
coulomb per kilogram	roentgens	x 3 876.0	*x 4 000*
coulomb per square centimeter	coulombs per square inch	x 6.451 6^	*x 6.5*
coulomb per square inch	coulombs per square centimeter	x 0.155 00	*x 0.16*
coulomb, absolute	coulombs, international	x 1.000 2	*x 1*
count (for fabrics)	number of warp yarns and weft yarns per inch	x 1^	
couple	similar or related things	x 2^	
cpi	characters per inch	x 1^	
cubic centimeter	milliliter	x 1^	
cubic decimeter	liter	x 1^	
cubic foot	bushels, struck measure	x 0.803 56	*x 0.8*
	cubic inches	x 1 728^	*x 1 700*
	cubic meters	x 0.028 317	*x 0.028*
	cubic yards	/27^ or x 0.037 037	*x 0.037*
	drams, fluid	x 7 660.1	*x 8 000*
	gallons	x 7.480 5	*x 7.5*
	gills	x 239.38	*x 240*
	liters	x 28.317	*x 28*
	milliliters	x 28 317	*x 28 000*
	minims	x 4.596 0 E+05	*x 4.6 E+05*
	ounces, fluid	x 957.51	*x 1 000*
	pecks	x 3.214 3	*x 3.2*
	pints, dry	x 51.428	*x 50*
	pints, fluid	x 59.844	*x 60*
	quarts, dry	x 25.714	*x 26*
	quarts, fluid	x 29.922	*x 30*
cubic foot per hour	acre-feet per hour	x 2.295 7 E-05	*x 2.3 E-05*
	barrels (42 gallons) per day	x 4.274 6	*x 4.3*
	barrels (42 gallons) per hour	x 0.178 11	*x 0.18*
	cubic feet per minute	/60^ or x 0.016 667	*x 0.017*
	cubic feet per second	/3 600^ or x 2.777 8 E-04	
	gallons per hour	x 7.480 5	*x 7.5*
	gallons per minute	x 0.124 68	*x 0.12*
	liters per second	x 0.007 865 8	*0.008*
cubic foot per minute	barrels (42 gallons) per day	x 256.47	*x 260*
	barrels (42 gallons) per hour	x 10.686	*x 11*
	cubic feet per hour	x 60^	
	cubic feet per second	/60^ or x 0.016 667	*x 0.017*
	cubic meters per second	x 4.719 5 E-04	*x 4.7 E-04*
	gallons per hour	x 448.83	*x 450*
	gallons per minute	x 7.480 5	*x 7.5*

TABLE 2
Measurement Conversions, Alphabetical

All measurement units are US, unless otherwise noted. All number denominations "billion" and higher are US, unless otherwise noted.

CONVERT	TO EQUIVALENT	BY PRECISELY	OR WITHIN ± 5.0 %
	liters per second	x 0.471 95	*x 0.47*
cubic foot per pound	cubic meters per kilogram	x 0.062 428	*/16 or x 0.06*
cubic foot per second	barrels (42 gallons) per day	x 15 388	*x 15 000*
	barrels (42 gallons) per hour	x 641.19	*x 640*
	cubic feet per hour	x 3 600^	
	cubic feet per minute	x 60^	
	cubic meters per second	x 0.028 317	*x 0.028*
	cubic yards per minute	x 20/9^ or x 2.222 2	*x 2.2*
	gallons per hour	x 26 930	*x 27 000*
	gallons per minute	x 448.83	*x 450 or x 4 000/9*
	liters per second	x 28.317	*x 28*
	million gallons per day	x 0.646 32	*x 0.65*
cubic foot (Can.)	cubic foot (US)	x 1^	
cubic inch	bushels, struck measure	x 4.650 3 E-04	*x 4.7 E-04*
	cubic feet	/1 728^ or x 5.787 0 E-04	*x 6 E-04*
	cubic meters	x 1.638 7 E-05	*x 1.6 E-05*
	cubic yards	x 2.143 3 E-05	*x 2.1 E-05*
	drams, fluid	x 4.432 9	*x 4.4*
	gallons	/231^ or x 0.004 329 0	*x 0.004 3*
	gills	x 0.138 53	*x 0.14*
	liters	x 0.016 387	*x 0.016*
	milliliters	x 16.387	*x 16*
	minims	x 265.97	*x 270*
	ounces, fluid	x 0.554 11	*x 0.55*
	pecks	x 0.001 860 1	*x 0.001 9*
	pints, dry	x 0.029 762	*x 0.03*
	pints, fluid	x 0.034 632	*x 0.035*
	quarts, dry	x 0.014 881	*x 0.015*
	quarts, fluid	x 0.017 316	*x 0.017*
cubic inch per minute	cubic meters per second	x 2.731 2 E-07	*x 2.7 E-07*
cubic inch (Can.)	cubic inch (US)	x 1^	
cubic meter	acre-feet	x 8.107 0 E-04	*x 0.11 E-04*
	barrels, petroleum (42 gallons)	x 6.289 8	*x 6.3*
	bushels, struck measure	x 28.378	*x 28*
	cubic feet	x 35.315	*x 35*
	cubic inches	x 61 024	*x 60 000*
	cubic yards	x 1.308 0	*x 1.3 or x 5/4*
	gallons	x 264.17	*x 260*
	liters	x 1 000^	
	milliliters	x 1^E+06	
	pecks	x 113.51	*x110*
	pints, dry	x 1 816.2	*x 1 800*
	quarts, dry	x 908.08	*x 900*
	stere	x 1^	
cubic meter per hour	gallons per minute	x 4.402 9	*x 40/9*
cubic meter per kilogram	cubic feet per pound	x 16.018	*x 16*
cubic meter per second	cubic feet per minute	x 2 118.9	*x 2 100*
	cubic feet per second	x 35.315	*x 35*
	cubic inches per minute	x 3.661 4 E+06	*x 3.7 E+06*
	cubic yards per minute	x 78.477	*x 80*
	gallons per day	x 2.282 4 E+07	*x 2.3 E+07*

TABLE 2
Measurement Conversions, Alphabetical

All measurement units are US, unless otherwise noted. All number denominations "billion" and higher are US, unless otherwise noted.

CONVERT	TO EQUIVALENT	BY PRECISELY	OR WITHIN ±5.0 %
	gallons per hour	x 9.510 2 E+05	*x 9.5 E+05*
	gallons per minute	x 15 850	*x 16 000*
cubic rod (Eng.)	cubic feet (Eng.)	x 1 000	
cubic yard	cubic feet	x 27^	
	cubic inches	x 46 656^	*x 47 000*
	cubic meters	x 0.764 55	*x 0.76*
	liters	x 764.55	*x 760*
	milliliters	x 7.645 5 E+05	*x 7.6 E+05*
cubic yard per minute	cubic feet per second	x 9/20^ or x 0.45^	*x 4/9*
	cubic meters per second	x 0.012 743	*/80 or x 0.013*
	gallons per second	x 3.366 2	*x 10/3 or x 3.4*
	liters per second	x 12.743	*x 25/2 or x 13*
cubic yard (Can.)	cubic yard (US)	x 1^	
cubit (ancient)	inches	x 17 to x 22 (approx.)	
cunit (for solid wood, Can.)	cubic feet	x 100^	
cup, measuring	milliliters	x 236.59	*x 240*
	ounces, fluid	x 8^	
	pints, fluid	/2^ or x 0.5^	
	tablespoons	x 16^	
	teaspoons	x 48^	
cup, measuring (Brit.)	ounces, Imperial fluid (Brit.)	x 10^	
cup, measuring (Can.)	ounces, Imperial fluid (Can.)	x 8^	
curie	becquerels	x 3.7^ E+10	
current density	ampere per square meter	x 1^	
cycle per second	hertz	x 1^	
cycle, metonic	lunar months	x 235	*x 240*
	years	x 19	
C_v (valve flow coefficient)	number of gallons per minute [7]	x 1^	
-D-			
daily	occurring every day	x 1^	
dalton	atomic mass unit	x 1^	
darcy (for oil fields)	square meters	x 9.869 2 E-13	*x 10 E-13*
	square micrometers	x 0.986 92	*x 1*
day	day, calendar	x 1^	
day (customary)	day, mean solar (midnight to midnight)	x 1^	
	hours	x 24^	
	minutes	x 1 440^	*x 1 400*
	seconds	x 86 400^	*x 86 000*
daylight-saving time	See "time, daylight-saving"		
day, calendar	day, mean solar	x 1^	
	hours, mean solar	x 24^	
	minutes, mean solar	x 1 440^	*x 1 400*
	seconds, mean solar	x 86 400^	*x 86 000*
day, civil	day	x 1^	
	hours	x 24^	
day, lunar	24 hr. 50 min. sidereal time	x 1	
day, mean equinoctial	day, mean sidereal	x 1^	

[7] C_v is the number of US gallons per minute of water at 60 deg. F flowing through a valve when the pressure drop across the valve is one pound per square inch under stated conditions of pressure and valve opening.

TABLE 2
Measurement Conversions, Alphabetical

All measurement units are US, unless otherwise noted. All number denominations "billion" and higher are US, unless otherwise noted.

CONVERT	TO EQUIVALENT	BY PRECISELY	OR WITHIN ± 5.0 %
day, mean solar	24 hr. 3 min. 56.56 sec. mean solar time	x 1	
	day (customary)	x 1^	
	days, sidereal	x 1.002 7	*x 1*
	day, calendar	x 1^	
	hours	x 24^	
	minutes	x 1 440^	*x 1 400*
	seconds	x 86 400^	*x 86 000*
day, natural	hours, inequal	x 12	
day, sidereal	23 hr. 56 min. 4.09 sec. mean solar time	x 1	
	day, mean solar	x 0.997 27	*x 1*
	hours, mean solar	x 23.934	*x 24*
	hours, sidereal	x 24^	
	minutes, mean solar	x 1 436.1	*x 1 400*
	minutes, sidereal	x 1 440^	*x 1 400*
	seconds, mean solar	x 86 164	*x 86 000*
	seconds, sidereal	x 86 400^	*x 86 000*
dB (with suffix *A, B, C,* or *D)*	decibels	x 1^	
dBA (most common) [15]	decibels on weighted *A* scale at 40 phons	x 1^	
dBB	decibels on weighted *B* scale at 70 phons	x 1^	
dBC	decibels on weighted *C* scale at 100 phons	x 1^	
dBD (for aircraft jet noise)	decibels on weighted *D* scale	x 1^	
decade (number)	ratio of 10 to 1	x 1^	
decade (quantity)	group of ten	x 1^	
decade (time)	years	x 10^	
decay time	time for a pulse or quantity to decline to a specified percent of initial value	x 1^	
decennary	years	x 10^	
decibel increase of one	sound-intensity increase factor	x cube root of 2 or x 1.259 9	*x 1.25*
decibel increase of three	sound-intensity increase factor	x 2	
decibel (for current)	ratio of two levels of current with equal resistance	x 10 to power {no. of decibels x 0.05}	
decibel (for electric power ratio)	nepers	x 0.115 13	*x 23/200 or /9*
	ratio of two electric power levels	x 10 to power {no. of decibels x 0.1}	
decibel (for sound-power-intensity level)	ratio of measured intensity to reference intensity [16]	x 10 to power {no. of decibels x 0.1}	
decibel (for sound-pressure-amplitude level)	ratio of measured pressure to reference pressure [17]	x 10 to power {no. of decibels x 0.05}	
decibel (for voltage)	ratio of two levels of voltage with equal resistance	x 10 to power {no. of decibels x 0.05}	
decillion (Brit.)	novemdecillion (US)	x 1^	
decillion (US)	1^E+33	x 1^	
decimal	based on the number 10	x 1^	
decimeter	feet	x 0.328 08	*x 0.33*
	inches	x 3.937 0	*x 4*
decuple	items	x 10^	
deep six	fathoms	x 6^	
degree (angle)	grades	x 10/9^ or x 1.111 1	*x 1.1*

[15] The dBA scale of sound-pressure level is weighted to approximate the equal-loudness contour curve of the human ear.
[16] The reference level for sound power is 1 E-16 watts per square centimeter.
[17] The reference level for sound pressure is 2 E-04 dynes per square centimeter.

TABLE 2
Measurement Conversions, Alphabetical

All measurement units are US, unless otherwise noted. All number denominations "billion" and higher are US, unless otherwise noted.

CONVERT	TO EQUIVALENT	BY PRECISELY	OR WITHIN ± 5.0 %
	mils (angle)	x 160/9^ or x 17.778	*x 18*
	minutes (angle)	x 60^	
	radians	x pi/180 or x 0.017 453	*x 0.017*
	revolutions	/360^ or x 0.002 777 8	*x 0.002 8*
	seconds (angle)	x 3 600^	
degree (angle) per second	revolutions per minute	/6^ or x 0.166 67	*x 0.17*
	revolutions per second	/360^ or x 0.002 777 8	*x 0.002 8*
deg. Barkometer (for tanning industry)	specific gravities above or below 1.000	x 1^	
deg. Celsius (span)	deg. Fahrenheit	x 9/5^ or x 1.8^	
	deg. Rankine	x 9/5^ or x 1.8^	
	deg. Reaumur	x 4/5^ or x 0.8^	
	kelvin	x 1^	
deg. Celsius (temp.)	deg. Fahrenheit	x 9/5^ + 32^	*x 1.8 + 32*
	deg. Rankine	x 9/5^ + 491.67	*x 1.8 + 500*
	deg. Reaumur	x 4/5^ or x 0.8^	
	kelvins	+ 273.15	*+ 270*
deg. centigrade (obsolete)	deg. Celsius	x 1^	
deg. Fahrenheit (span)	deg. Celsius	x 5/9^ or x 0.555 56	*x 0.56*
	deg. Rankine	x 1^	
	deg. Reaumur	x 4/9^ or x 0.444 44	*x 0.44*
	kelvins	x 5/9^ or x 0.555 56	*x 0.56*
deg. Fahrenheit (temp.)	deg. Celsius	- 32^ x 5/9^	*- 32 x 0.56*
	deg. Rankine	+ 459.67	*+ 460*
	deg. Reaumur	- 32^ x 4/9^	*- 32 x 0.44*
	kelvins	+ 459.67 x 5/9^	*+ 460 x 0.56*
deg. Quevenne (for milk)	specific gravities above 1.000	x 1^	
deg. Rankine (span)	deg. Celsius	x 5/9^ or 0.555 56	*x 0.56*
	deg. Fahrenheit	x 1^	
	kelvins	x 5/9^ or 0.555 56	*x 0.56*
deg. Rankine (temp.)	deg. Celsius	- 491.67 x 5/9^	*-500 x 0.56*
	deg. Fahrenheit	- 459.67	*- 460*
	kelvins	x 5/9^ or 0.555 56	*x 0.56*
deg. Reaumur (span)	deg. Celsius	x 5/4^ or x 1.25^	*x 1.3*
	deg. Fahrenheit	x 9/4^ or x 2.25^	*x 2.3*
deg. Reaumur (temp.)	deg. Celsius	x 5/4^ or x 1.25^	*x 1.3*
	deg. Fahrenheit	x 9/4^ + 32^	*x 2.3 + 32*
deg. Twaddell (for heavier-than-water liquids)	specific gravities above 1.000	x 1^	
dekaliter	gallons	x 2.641 7	*x 2.6*
dekameter	rods (length)	x 1.988 4	*x 2*
dekatherm	Btu	x 1^E+06	
demiard (Can.)	liters	x 0.284 13	*x 0.28*
demijohn	gallons	x 1 to x 10	
denier (for thread thickness)	grams per meter	/9 000^ or x 1.111 1 E-04	*x 1.1 E-04*
denier (for yarns)	milligrams per meter	/9^ or x 0.111 11	*x 0.11*
density	1/specific volume	x 1^	
density of water (@ 104 deg. F)	pounds per cubic foot	x 61.942	*x 60*
	pounds per gallon	x 8.280 4	*x 8*

TABLE 2
Measurement Conversions, Alphabetical

All measurement units are US, unless otherwise noted. All number denominations "billion" and higher are US, unless otherwise noted.

CONVERT	TO EQUIVALENT	BY PRECISELY	OR WITHIN ± 5.0 %
density of water (@ 140 deg. F)	pounds per cubic foot	x 61.390	*x 60*
	pounds per gallon	x 8.206 7	*x 8*
density of water (@ 4 deg. C)	pounds per cubic foot	x 62.426	*x 60*
	pounds per gallon	x 8.345 2	*x 8*
density of water (@ 68 deg. F)	pounds per cubic foot	x 62.316	*x 60*
	pounds per gallon	x 8.330 4	*x 8*
density (optical)	log of 1/transmittance	x 1	
density, relative	specific gravity	x 1^	
dibit	a binary-number arrangement: 00, 01, 10, or 11	x 1^	
diopters (optical power of lens)	1/centimeters (of focal length)	x 100^	
	1/meters (of focal length)	x 1^	
disintegration per second (activity of radionuclide)	becquerel	x 1^	
diurnal	daily	x 1^	
dose equivalent	sievert	x 1^	
double	twice as large	x 1	
dozen	items	x 12^	
dozen, baker's	items	x 13^	
dpi	dots per inch	x 1^	
drachm (Brit.)	ounce, apothecary	/8^ or x 0.125^	*x 0.13*
drachm, apothecary (Brit.)	dram, apothecary (US)	x 1^	
drachm, fluid apothecary (Brit.)	drachm, Imperial fluid (Brit.)	x 1^	
drachm, Imperial fluid (Brit.)	cubic inches	x 0.216 73	*x 0.22*
	drams, fluid (US)	x 0.960 76	*x 1*
	milliliters	x 3.551 6	*x 3.6*
	minims, Imperial (Brit.)	x 60^	
	scruples, Imperial fluid (Brit.)	x 3^	
dram, apothecary	drams, avoirdupois	x 384/175^ or x 2.194 3	*x 2.2*
	grains	x 60^	
	grams	x 3.887 9	*x 4*
	kilograms	x 0.003 887 9	*x 0.004*
	milligrams	x 3 887.9	*x 4 000*
	ounces, apothecary	/8^ or x 0.125^	*x 0.13*
	ounces, avoirdupois	x 24/175^ or x 0.137 14	*x 0.14*
	ounces, troy	/8^ or x 0.125^	*x 0.13*
	pennyweights	x 2.5^	
	pounds, advoirdupois	x 3/350^ or x 0.008 571 4	*x 0.008 6*
	pounds, apothecary	/96^ or x 0.010 417	*x 0.01*
	pounds, troy	/96^ or x 0.010 417	*x 0.01*
	scruples, apothecary	x 3^	
dram, avoirdupois	drams, apothecary	x 175/384^ or x 0.455 73	*x 0.46*
	grains	x 875/32^ or x 27.344	*x 27*
	grams	x 1.771 8	*x 1.8*
	kilograms	x 0.001 771 8	*x 0.001 8*
	milligrams	x 1 771.8	*x 1 800*
	ounces, apothecary	x 175/3 072^ or x 0.056 966	*x 0.057*
	ounces, avoirdupois	/16^ or x 0.062 5^	*x 0.063*

TABLE 2
Measurement Conversions, Alphabetical

All measurement units are US, unless otherwise noted. All number denominations "billion" and higher are US, unless otherwise noted.

CONVERT	TO EQUIVALENT	BY PRECISELY	OR WITHIN ± 5.0 %
	ounces, troy	x 175/3 072^ or x 0.056 966	*x 0.057*
	pennyweights	x 875/768^ or x 1.139 3	*x 1.1*
	pounds, apothecary	x 175/36 864^ or x 0.004 747 2	*x 0.004 7*
	pounds, avoirdupois	/256^ or x 0.003 906 3	*x 0.004*
	pounds, troy	x 175/36 864^ or x 0.004 747 2	*x 0.004 7*
	scruples, apothecary	x 175/128^ or x 1.367 2	*x 1.4*
dram, avoirdupois (Brit.)	dram, avoirdupois (US)	x 1^	
dram, avoirdupois (Can.)	dram, avoirdupois (US)	x 1^	
dram, fluid	cubic feet	x 1.305 5 E-04	*x 1.3 E-04*
	cubic inches	x 0.225 59	*x 0.23*
	gallons	/1 024^ or x 9.765 6 E-04	*x 0.001*
	gills	/32^ or x 0.031 25^	*x 0.03*
	liters	x 0.003 696 7	*x 0.003 7*
	milliliters	x 3.696 7	*x 3.7*
	minims	x 60^	
	ounces, fluid	/8^ or x 0.125^	*x 0.13*
	pints, fluid	/128^ or x 0.007 812 5^	*x 0.008*
	quarts, fluid	/256^ or x 0.003 906 3	*x 0.004*
dram, fluid apothecary	dram, fluid	x 1^	
dram, fluid (Can.)	drachm, Imperial fluid (Brit.)	x 1^	
dram, fluid (US)	drachms, Imperial fluid (Brit.)	x 1.040 8	*x 1*
drop (Can.)	milliliter	/20^ or x 0.05^	
	teaspoon	/100^ or x 0.01^	
duet	group of two	x 1^	
duo	group of two	x 1^	
duodecillion (Brit.)	1^E+72	x 1^	
duodecillion (US)	1^E+39	x 1^	
duodecimal	based on the number 12	x 1^	
duodenary	based on the number 12	x 1^	
duosexadecimal	duotricinary	x 1^	
duotricinary	based on the number 32	x 1^	
dyad	group of two	x 1^	
dyne	gram-centimeter per second per second	x 1^	
	joule per centimeter	x 1^E-07	
	joules per meter	x 1^E-05	
	newtons	x 1^E-05	
	pounds (force)	x 2.248 1 E-06	*x 2.2 E-06*
dyne per square centimeter	bars	x 1^E-06	
	pascals	/10^ or x 0.1^	
	pounds (force) per square inch	x 1.450 4 E-05	*x 1.5 E-05*
	torrs	x 7.500 6 E-04	*x 7.5 E-04*
dyne-centimeter	foot-pounds (force)	x 7.375 6 E-08	*x 7.5 E-08*
	meter-kilograms (force)	x 1.019 7 E-08	*x 1 E-08*
	newton-meters	x 1^E-07	
-E-			
e	2.718 28	x 1	*2.7*
Earth ice-age cycle	years	x 100 000 (approx.)	
einstein	mole of photons	x 1^	
electric charge density	coulomb per cubic meter	x 1^	
electric field strength	volt per meter	x 1^	

TABLE 2
Measurement Conversions, Alphabetical

All measurement units are US, unless otherwise noted. All number denominations "billion" and higher are US, unless otherwise noted.

CONVERT	TO EQUIVALENT	BY PRECISELY	OR WITHIN ± 5.0 %
electric flux density	coulomb per square meter	x 1^	
electromagnetic spectrum	See APPENDIX, "electromagnetic spectrum"		
electron-volt	atomic mass units (equivalent mass)	x 6.022 5 E+23	*x 6 E+23*
	ergs	x 1.602 2 E-12	*x 1.6 E-12*
elite (typewriter type)	characters per inch	x 12^	
ell (Eng.)	centimeters	x 114.3^	*x 110*
	inches	x 45^	
em (printer's)	ens (printer's)	x 2^	
	inches	x 0.166 04	*x 0.17*
	points (printer's)	x 12^	
EMU of capacitance	farads	x 1^E+09	
EMU of current	amperes	x 10^	
EMU of electric potential	volts	x 1^E-08	
EMU of inductance	henrys	x 1^E-09	
EMU of resistance	ohms	x 1^E-09	
em, pica	centimeters	x 0.421 75	*x 0.42*
	inches	x 0.166 04	*x 0.17*
en (printer's)	ems (printer's)	/2^	
energy density	joule per cubic meter	x 1^	
entropy	joule per kelvin	x 1^	
entropy, specific	ratio of entropy of a substance to its mass	x 1^	
erg	electron-volts	x 6.241 5 E+11	*x 6 E+11*
	grams (equivalent mass)	x 1.112 6 E-21	*x 1.1 E-21*
	joules	x 1^E-07	
erg per second	watts	x 1^E-07	
erg per second-square centimeter	watts per square meter	/1 000^ or x 0.001^	
ESU of capacitance	farads	x 1.112 7 E-12	*x 1.1 E-12*
ESU of current	amperes	x 3.335 6 E-10	*x 3.3 E-10*
ESU of electric potential	volts	x 299.79	*x 300*
ESU of inductance	henrys	x 8.987 6 E+11	*x 9 E+11*
ESU of resistance	ohms	x 8.987 6 E+11	*x 9 E+11*
-F-			
farad	abfarads	x 1^E-09	
	EMU of capacitance	x 1^E-09	
	ESU of capacitance	x 8.987 6 E+11	*x 9 E+11*
	statfarads	x 8.987 6 E+11	*x 9 E+11*
faraday (based on carbon-12)	abcoulombs	x 9 648.7	*x 10 000*
	ampere-hours	x 26.302	*x 26*
	coulombs	x 96 487	*x 100 000*
	statcoulombs	x 2.892 6 E+14	*x 3 E+14*
faraday, chemical	coulombs	x 96 496	*x 100 000*
faraday, physical	ampere-hours	x 26.812	*x 27*
	coulombs	x 96 522	*x 100 000*
farad, absolute	farads, international	x 1.000 5	*x 1*
fathom	feet	x 6^	
	meters	x 1.828 8	*x 1.8*
fermi	meters	x 1^E-15	
fifth (of liquor)	gallons	/5^ or x 0.2^	

TABLE 2
Measurement Conversions, Alphabetical

All measurement units are US, unless otherwise noted. All number denominations "billion" and higher are US, unless otherwise noted.

CONVERT	TO EQUIVALENT	BY PRECISELY	OR WITHIN ± 5.0 %
	quarts, fluid	x 4/5^ or x 0.8^	
fineness (for precious-metal alloys)	parts per thousand (by weight)	x 1^	
fineness (of gold or silver)	parts per thousand (by weight)	x 1^	
firkin	gallons	x 9	
firkin (Brit.)	barrels, Imperial (Brit.)	/4 or x 0.25 (usually)	
	gallons, ale (Brit.)	x 8	
	gallons, Imperial (Brit.)	x 9	
firkin (of butter) (Brit.)	pounds	x 56^	
flagon	quarts, fluid	x 2 (usually)	
flank speed (for ships)	maximum-capable speed of ship	x 1	
flask (of mercury)	pounds	x 76^	
floating-point operations per second	calculations per second	x 1^	
flop	floating-point operation	x 1^	
flops	floating-point operations per second	x 1^	
food	See "gram of ..." and "ounce of ..."		
foot	centimeters	x 30.48^	*x 30*
	chains, engineer's	/100^ or x 0.01^	
	chains, surveyor's	/66^ or x 0.015 152	*x 0.015*
	fathoms	/6^ or x 0.166 67	*x 0.17*
	foot, statute	x 1^	
	foot, survey	x 1	
	furlongs	/660^ or x 0.001 515 2	*x 0.001 5*
	inches	x 12^	
	links, engineer's	x 1^	
	links, surveyor's	/0.66^ or x 1.515 2	*x 1.5*
	meters	x 0.304 8^	*x 0.3*
	miles	/5 280^ or x 1.893 9 E-04	*x 1.9 E-04*
	miles, nautical (intl.)	x 1.645 8 E-04	*x 1.6 E-04*
	mils (length)	x 12 000^	
	perch (length)	/16.5^ or x 0.060 606	*x 0.06*
	rods (length)	/16.5^ or x 0.060 606	*x 0.06*
	yards	/3^ or x 0.333 33	*x 0.33*
foot of water (@ 60 deg. F)	pounds (force) per square inch	x 0.433 09	*x 0.43*
foot of water [13]	pascals	x 2 989.0	*x 3 000*
foot per 100 feet	grade (angle), percent	x 1^	
foot per hour	meters per second	x 8.466 7 E-05	*x 8.5 E-05*
foot per minute	meters per second	x 0.005 08^	*x 0.005*
foot per second	meters per second	x 0.304 8^	*x 0.3*
foot per second per second	centimeters per second per second	x 30.48^	*x 30*
	meters per second per second	x 0.304 8^	*x 0.3*
foot (Brit.)	foot (US)	x 1^	
foot (Can.)	foot (US)	x 1^	
foot (intl.)	foot, statute (US)	x 1	
	foot, survey (US)	x 1	
	meters	x 0.304 8^	*x 0.3*
foot, board	See "board foot"		

[13] Unless otherwise noted, liquid-head conversions are based on: a pressure of one standard atmosphere; temperature for mercury = 0.0 deg. C = 32.0 deg. F; temperature for water = 4.0 deg. C = 39.2 deg. F.

TABLE 2
Measurement Conversions, Alphabetical

All measurement units are US, unless otherwise noted. All number denominations "billion" and higher are US, unless otherwise noted.

CONVERT	TO EQUIVALENT	BY PRECISELY	OR WITHIN ± 5.0 %
foot, mil	See "mil-foot"		
foot, Paris (French land length, Queb.)	foot (French measure, Queb.)	x 1^	
	inches (English measure, Queb.)	x 12.789^	*x 13*
foot, solid	cubic foot	x 1^	
foot, statute	meters	x 0.304 8^	*x 0.3*
foot, superficial	square foot	x 1^	
foot, survey (for land measure) [9]	meters	x 1 200/3 937^ or x 0.304 80	*x 0.3*
foot-candle	lumen per square foot	x 1^	
	lumens per square meter	x 10.764	*x 11*
	lux	x 10.764	*x 11*
foot-candle, apparent	foot-lambert	x 1^	
foot-lambert	candelas per square foot	x 0.318 31	*x 0.32*
	candelas per square meter	x 3.426 3	*x 3.4*
	lumen per square foot	x 1^	
foot-pound (force)	Btu (IT)	x 0.001 285 1	*x 0.001 3*
	calories (IT)	x 0.323 83	*x 0.32*
	horsepower-hours	x 500 E-07/99^ or x 5.050 5 E-07	*x 5 E-07*
	horsepower-hours, metric	x 5.120 6 E-07	*x 5 E-07*
	joules	x 1.355 8	*x 1.4*
	kilocalories	x 3.238 3 E-04	*x 3.2 E-04*
	kilowatt-hours	x 3.766 2 E-07	*x 3.8 E-07*
	liter-atmospheres	x 0.013 381	*x 0.013*
	meter-kilograms (force)	x 0.138 25 x 0.138 25	*x 0.14*
foot-pound (force) per hour	watts	x 3.766 2 E-04	*x 3.8 E-04*
foot-pound (force) per minute	watts	x 0.022 597	*x 0.023*
foot-pound (force) per second	horsepower	/550^ or x 0.001 818 2	*x 0.001 8*
	horsepower, metric	x 0.001 843 4	*x 0.001 8*
	kilocalories (IT) per second	x 3.238 3 E-04	*x 3.2 E-04*
	meter-kilograms (force) per second	x 0.138 25	*x 0.14*
	watts	x 1.355 8	*x 1.4*
foot-poundal	joules	x 0.042 140	*x 0.042*
force of gravity	weight	x 1^	
fortnight	weeks	x 2^	
forty	acres	x 40^	
	quarter sections	/4^ or x 0.25^	
FPP tornado scale	See APPENDIX, "wind speeds"		
freezing point of water (ITS)	deg. Celsius (@ 101.325 kPa)	x 0^	
	deg. Fahrenheit (@ 14.696 psia)	x 32^	
frequency	1/period	x 1^	
frequency for musical tone *A*, standard (intl.)	hertz	x 440.0 ± 0.5	
frequency ratio for half-tone intervals (for tempered musical scale)	the 12th root of 2 (= 1.059 5)	x 1	

[9] The length of the survey foot may be revised.

TABLE 2
Measurement Conversions, Alphabetical

All measurement units are US, unless otherwise noted. All number denominations "billion" and higher are US, unless otherwise noted.

CONVERT	TO EQUIVALENT	BY PRECISELY	OR WITHIN ± 5.0 %
frequency ratio for whole-tone intervals (for tempered musical scale)	the 6th root of 2 (= 1.122 5)	x 1	
frigorie (for refrigeration, European)	Btu per minute	x 50 (approximate)	
Fujita intensity scale (for winds)	See APPENDIX, "wind speeds"		
furlong	feet	x 660^	
	meters	x 201.17	*x 200*
	miles	/8^ or x 0.125^	*x 0.13*
	rods	x 40^	
furlong (Can.)	furlong (US)	x 1^	
f-number	ratio of lens focal length to aperture effective diameter	x 1^	
f-stop	f-number	x 1^	
-G-			
g	See "gravity, acceleration of"	x 1^	
gal (for geodesy)	centimeters per second per second	x 1^	
	meters per second per second	/100^ or x 0.01^	
gallon	cubic feet	x 0.133 68	*x 0.13*
	cubic inches	x 231^	*x 230*
	cubic meters	x 0.003 785 4	*x 0.003 8*
	dekaliters	x 0.378 54	*x 0.38*
	drams, fluid	x 1 024^	*x 1 000*
	gallons, Imperial (Brit.)	x 0.832 67	*x 0.8*
	gallon, fluid	x 1^	
	gallons, dry	x 0.859 37	*x 0.9*
	gills	x 32^	
	hectoliters	x 0.037 854	*x 0.038*
	liters	x 3.785 4	*x 3.8*
	milliliters	x 3 785.4	*x 3 800*
	minims	x 61 440^	*x 60 000*
	ounces, fluid	x 128^	*x 130*
	pints, fluid	x 8^	
	quarts, fluid	x 4^	
gallon per day	cubic meters per second	x 4.381 3 E-08	*x 4.4 E-08 or x 4/9 E-08*
gallon per hour	barrels (42 gallons) per day	x 0.571 43	*x 0.57*
	barrels (42 gallons) per hour	x 0.023 810	*x 0.024*
	cubic feet per hour	x 0.133 68	*x 0.13*
	cubic feet per minute	x 0.002 228 0	*x 0.002 2*
	cubic feet per second	x 3.713 3 E-05	*x 3.7 E-05*
	cubic meters per second	x 1.051 5 E-06	*x 1.1 E-06*
	gallons per minute	/60^ or x 0.016 667	*x 0.017*
	liters per second	x 0.001 051 5	*x 0.001 1*
gallon per minute	acre-feet per hour	x 1.841 3 E-04	*x 1.8 E-04*
	barrels (42 gallons) per day	x 34.286	*x 34*
	barrels (42 gallons) per hour	x 1.428 6	*x 1.4*
	cubic feet per hour	x 8.020 8	*x 8*
	cubic feet per minute	x 0.133 68	*x 0.13*
	cubic feet per second	x 0.002 228 0	*x 0.002 2*
	cubic meters per second	x 6.309 0 E-05	*x 6.3 E-05*

TABLE 2
Measurement Conversions, Alphabetical

All measurement units are US, unless otherwise noted. All number denominations "billion" and higher are US, unless otherwise noted.

CONVERT	TO EQUIVALENT	BY PRECISELY	OR WITHIN ± 5.0 %
	gallons per hour	x 60^	
	liters per minute	x 3.785 4	*x 3.8*
	liters per second	x 0.063 090	*x 0.063*
	tons of water per day	x 6.008 6 (@ 4 deg Celsius)	*x 6*
gallon (Can.)	gallon, Imperial (Brit.)	x 1^	
gallonage	number of gallons	x 1	
gallon, ale (Brit.)	liters	x 4.62	*x 4.6*
gallon, beer (Brit.)	gallon, Imperial ale (Brit.)	x 1^	
gallon, dry (US, not legal)	bushel	x 1/8^ or x 0.125^	*x 0.13*
	cubic inches	x 268.80	*x 270*
	gallons, fluid (US)	x 1.163 6	*x 1.2*
gallon, fluid (Can.)	cubic meters	x 0.004 546 1	*x 0.004 5*
	gallons, fluid (US)	x 1.201 0	*x 1.2*
	gallon, Imperial (Brit.)	x 1^	
gallon, Imperial apothecary (Brit.)	gallon, Imperial (Brit.)	x 1^	
gallon, Imperial (Brit.)	bushels, Imperial (Brit.)	/8^ or x 0.125^	*x 0.13*
	cubic inches	x 277.42	*x 280*
	cubic meters	x 0.004 546 1	*x 0.004 5*
	gallons, fluid (US)	x 1.201 0	*x 1.2*
	liters	x 4.546 1	*x 4.5*
	ounces, Imperial fluid (Brit.)	x 160^	
	pints, Imperial (Brit.)	x 8^	
	quarts, Imperial (Brit.)	x 4^	
gallon, wine (Brit.)	cubic inches	x 231^	*x 230*
gamma (magnetic)	gauss	x 1^E-05	
	line per square centimeter	x 1^	
	lines per square inch	x 6.451 6^	*x 6.5*
	maxwell per square centimeter	x 1^	
	oersteds	x 1^E-05	
	teslas	x 1^E-09	
	webers per square centimeter	x 1^E-08	
	webers per square inch	x 6.451 6^ E-08	*x 6.5 E-08*
	webers per square meter	x 1^E-04	
gamma (mass)	grams	x 1^E-06	
	microgram	x 1^	
gauss (magnetic)	teslas	x 1^E-04	
GB	gigabyte	x 1^	
geepound	slug	x 1^	
geologic time	See "time scale, geologic"		
gestation period, human average	days	x 267	
gestation period, human range	days	x 250 to x 290	
Gflops	gigaflops	x 1^	
gigabyte	bytes	x 2^ to power 30 or x 1.073 7 E+09	*x billion*
gigaflops	billion (US) floating-point operations per second	x 1	
gilbert	abampere-turns	/4 pi or x 0.079 577	*x 0.08*
	amperes	x 0.795 77	*x 0.8*
	ampere-turns	/0.4 pi or x 0.795 77	*x 0.8*
gilbert per centimeter	ampere-turns per inch	x 2.021 3	*x 2*
	ampere-turns per meter	x 79.577	*x 80*
	oersted	x 1^	

TABLE 2
Measurement Conversions, Alphabetical

All measurement units are US, unless otherwise noted. All number denominations "billion" and higher are US, unless otherwise noted.

CONVERT	TO EQUIVALENT	BY PRECISELY	OR WITHIN ± 5.0 %
gill	cubic feet	x 0.004 177 5	*x 0.004*
	cubic inches	x 7.218 8	*x 7*
	drams, fluid	x 32^	
	gallons	/32^ or x 0.031 25^	*x 0.03*
	liters	x 0.118 29	*x 0.12*
	milliliters	x 118.29	*x 120*
	minims	x 1 920^	*x 2 000*
	ounces, fluid	x 4^	
	pints, fluid	/4^ or x 0.25^	
	quarts, fluid	/8^ or x 0.125^	*x 0.13*
gill (Brit.)	cubic meters	x 1.420 7 E-04	*x 1.4 E-04*
gill (Can.)	gill, Imperial (Brit.)	x 1^	
gill, Imperial (Brit.)	pints, Imperial (Brit.)	/4^ or x 0.25^	
golden ratio	See "ratio, golden"		
golden section	See "ratio, golden"		
googol	10^E+100	x 1^	
googolplex	10^ to the power googol	x 1^	
gpd	gallon per day	x 1^	
gph	gallon per hour	x 1^	
gpm	gallon per minute	x 1^	
gps	gallon per second	x 1^	
grad	See "grade"		
grade	degrees (angle)	x 0.9^	
	radians	x pi/200 or x 0.015 708	*x 0.016*
grain	carats	x 0.323 99	*x 0.32*
	drams, apothecary	/60^ or x 0.016 667	*x 0.017*
	drams, avoirdupois	x 0.036 571	*x 0.037*
	grams	x 0.064 799	*x 0.065*
	kilograms	x 6.479 9 E-05	*x 6.5 E-05*
	milligrams	x 64.799	*x 65*
	ounces, apothecary	/480^ or x 0.002 083 3	*x 0.002*
	ounces, avoirdupois	2/875^ or x 0.002 285 7	*x 0.002 3*
	ounces, troy	/480^ or x 0.002 083 3	*x 0.002*
	pennyweights	/24^ or x 0.041 667	*x 0.04*
	pounds, apothecary	/5 760^ or x 1.736 1 E-04	*x 1.7 E-04*
	pounds, avoirdupois	/7 000^ or x 1.428 6 E-04	*x 1.4 E-04*
	pounds, troy	/5 760^ or x 1.736 1 E-04	*x 1.7 E-04*
	scruples, apothecary	/20^ or x 0.05^	
grain per cubic foot	milligrams per cubic meter	x 2 288.4	*x 2 200*
	pounds per thousand cubic feet	/7^ or x 0.142 86	*x 0.14*
grain per gallon	pounds per gallon	/7 000^ or x 1.428 6 E-04	*x 1.4 E-04*
	grains per Imperial gallon (Brit.)	x 1.201 0	*x 1.2*
	grams per cubic meter	x 17.118	*x 17*
	kilograms per cubic meter	x 0.017 118	*x 0.017*
	pounds per million gallons	x 1 000/7^ or x 142.86	*x 140*
grain per gallon (of water @ 20 deg. C)	parts per million (by weight)	x 17.149	*x 17*
grain per gallon (of water @ 4 deg. C)	parts per million (by weight)	x 17.118	*x 17*
grain per Imperial gallon (Brit.)	pounds per Imperial gallon (Brit.)	/7 000^ or x 1.428 6 E-04	*x 1.4 E-04*
	pounds per million Imperial gallons (Brit.)	x 1 000/7^ or x 142.86	*x 140*
grain per Imperial gallon (of water @ 20 deg. C, Brit.)	grains per gallon (US)	x 0.832 67	*x 0.8*

TABLE 2
Measurement Conversions, Alphabetical

All measurement units are US, unless otherwise noted. All number denominations "billion" and higher are US, unless otherwise noted.

CONVERT	TO EQUIVALENT	BY PRECISELY	OR WITHIN ±5.0 %
	parts per million (by weight)	x 14.279	x 14
grain per Imperial gallon (of water @ 4 deg. C, Brit.)	parts per million (by weight)	x 14.254	x 14
grain (Can.)	grain (US)	x 1^	
grain, apothecary	grain, avoirdupois	x 1^	
	grain, troy	x 1^	
grain, apothecary (Brit.)	grain, apothecary (US)	x 1^	
grain, avoirdupois	grain, apothecary	x 1^	
	grain, troy	x 1^	
grain, avoirdupois (Brit.)	grain, avoirdupois (US)	x 1^	
grain, carat	See "carat grain"		
grain, pearl	See "pearl grain"		
grain, troy	grain, apothecary	x 1^	
	grain, avoirdupois	x 1^	
grain, troy (Brit.)	grain, troy (US)	x 1^	
gram	drams, apothecary	x 0.257 21	x 0.26
	drams, avoirdupois	x 0.564 38	x 0.56
	ergs (equivalent energy)	x 8.987 6 E+20	x 9 E+20
	grains	x 15.432	x 15
	joules (equivalent energy)	x 8.987 6 E+13	x 9 E+13
	kilograms	/1 000^ or x 0.001^	
	kilowatt-hours (equivalent energy)	x 2.496 5 E+10	
	milligrams	x 1 000^	
	ounces, apothecary	x 0.032 151	x 0.032
	ounces, avoirdupois	x 0.035 274	x 0.035
	ounces, troy	x 0.032 151	x 0.032
	pennyweights	x 0.643 01	x 0.64
	pounds, apothecary	x 0.002 679 2	x 0.002 7
	pounds, avoirdupois	x 0.002 204 6	x 0.002 2
	pounds, troy	x 0.002 679 2	x 0.002 7
	scruples, apothecary	x 0.771 62	x 0.8
gram of carbohydrate	kilocalories [5]	x 4	
gram of fat	kilocalories [5]	x 9	
gram of protein	kilocalories [5]	x 4	
gram per cubic centimeter	pounds per cubic foot	x 62.428	x 60
	pounds per cubic inch	x 0.036 127	x 0.036
gram per cubic meter	grains per cubic foot	x 0.437 00	x 4/9 or x 0.44
	grains per gallon	x 0.058 418	x 0.06
	milligram per liter	x 1^	
gram per liter	grains per gallon	x 58.418	x 60
	kilogram per cubic meter	x 1^	
	pounds per 1 000 gallons	x 8.345 4	x 8
	pounds per cubic foot	x 0.062 428	x 0.06
gram per liter (of water @ 4 deg. C)	parts per million (by weight)	x 1 000^	
gram per meter	ounces per foot	x 0.010 752	x 0.01
gram per milliliter	kilograms per cubic meter	x 1 000^	
	pounds per cubic foot	x 62.428	x 60

[5] The energy value of food and drink is customarily stated in "calories", but the technically correct measuring unit is "kilocaries", which is sometimes called "large calories". (1 kilocalorie = 1 000 calories) A person on a reducing diet may take 1 100 "calories" per day, but, in truth, he is taking 1 100 kilocalories, or 1.1 million (1 100 000) calories, or 1.1 megacalories.

TABLE 2
Measurement Conversions, Alphabetical

All measurement units are US, unless otherwise noted. All number denominations "billion" and higher are US, unless otherwise noted.

CONVERT	TO EQUIVALENT	BY PRECISELY	OR WITHIN ± 5.0 %
	pounds per cubic inch	x 0.036 127	*x 0.036*
	pounds per gallon	x 8.345 4	*x 8*
	tons, short per cubic yard	x 0.842 78	*x 0.84*
gram (force)	dynes	x 980.67	*x 1 000*
	newtons	x 9 806.7	*x 10 000*
	poundals	x 0.070 932	*x 0.07*
gram (force) per square centimeter	pascals	x 98.067	*x 100*
	pounds (force) per square foot	x 2.048 2	*x 2*
	pounds (force) per square inch	x 0.014 223	*x 0.014*
gram (force)-centimeter	dyne-centimeters	x 980.67	*x 1 000*
	foot-pounds (force)	x 7.233 0 E-05	*x 7 E-05*
gramme	gram (US)	x 1^	
gram-atom	chemical-element mass, in grams, equal in number to atomic weight	x 1^	
gram-atomic weight	gram-atom	x 1^	
gram-molecular weight	mole	x 1^	
gram-molecule	mole	x 1^	
gram-square centimeter	kilogram-square meters	x 1^E-07	
	pound-square feet	x 2.373 0 E-06	*x 2.4 E-06*
	pound-square inches	x 3.417 2 E-04	*x 3.4 E-04*
	slug-square feet	x 7.375 6 E-08	*x 7.4 E-08*
gravity, acceleration of[8]	centimeters per second per second	x 980.665^	*x 1 000*
	feet per second per second	x 32.174 0	*x 32*
	meters per second per second	x 9.806 65^	*x 10*
gray	rads (absorbed dose)	x 100^	
gross	dozens	x 12^	
	items	x 144^	
-H-			
half-life	time period for the potency or quantity of a substance to be reduced by one half	x 1^	
hand (for height of horses)	centimeters	x 10.16^	*x 10*
	inches	x 4^	
heat capacity	joule per kelvin	x 1^	
heat-flux density	watt per square meter	x 1	
hectare	acres	x 2.471 0	*x 2.5*
	square chains, surveyor's	x 24.710	*x 25*
	square feet	x 1.076 4 E+05	*x 1.1 E+05*
	square hectometer	x 1^	
	square meters	x 10 000^	
	square miles	x 0.003 861 0	*x 3/800 or x 0.004*
	square rods	x 395.37	*x 400*
hectare-meter	cubic meters	x 10 000^	
hectoliter	bushels, struck measure	x 2.837 8	*x 2.8*
	gallons	x 26.417	*x 26*
hectopascal	millibar	x 1^	
hefner	10-candlepower pentane candle	x 0.090	
	candela	x 0.92	*x 0.9*

8 Standard conditions for the acceleration of gravity are: 9.806 65 meters per second per second = 32.174 0 feet per second per second, at sea level and latitude 45 degrees. At other locations, the acceleration may differ within a span of more than 0.5 percent of the standard value.

TABLE 2
Measurement Conversions, Alphabetical

All measurement units are US, unless otherwise noted. All number denominations "billion" and higher are US, unless otherwise noted.

CONVERT	TO EQUIVALENT	BY PRECISELY	OR WITHIN ± 5.0 %
	candle, international	x 0.90	
	carcel	x 0.094	
	English candle	x 0.864	*x 0.9*
hemisphere	spheres	/2^ or x 0.5^	
	spherical right angles	x 4^	
	steradians	x 2 pi or x 6.283 2	*x 6.3*
henry	abhenrys	x 1^E+09	
	EMU of inductance	x 1^E+09	
	ESU of inductance	x 1.112 6 E-12	*x 1.1 E-12*
	stathenrys	x 1.112 6 E-12	*x 1.1 E-12*
henry, absolute	henrys, international	x 0.999 51	*x 1*
heptad	group of seven	x 1^	
hertz	1/second	x 1^	
	cycle per second	x 1^	
hexad	group of six	x 1^	
hexadecimal	sexadecimal	x 1^	
hide (Eng.)	acres	x 120	
hogshead	gallons	x 62.5 to x 140	
	gallons	x 63 (usually)	
	liters	x 238.48	*x 240*
hogshead (Brit.)	gallons (US)	x 64.851	*x 65*
	gallons, Imperial (Brit.)	x 54	
hogshead (Can.)	liters	x 245.49	*x 250*
horsepower	foot-pounds (force) per minute	x 33 000^	
	foot-pounds (force) per second	x 550^	
	horsepower, metric	x 1.013 9	*x 1*
	kilocalories (IT) per minute	x 10.686	*x 11*
	kilocalories (IT) per second	x 0.178 11	*x 0.18*
	kilocalories (thermochemical) per minute	x 10.694	*x 11*
	kilocalories (thermochemical) per second	x 0.178 23	*x 0.18*
	kilowatts	x 0.745 70	*x 0.75*
	meter-kilograms (force) per second	x 76.040	*x 76*
	watts	x 745.70	*x 750*
horsepower, boiler	Btu (IT) per hour	x 33 471	*x 33 000*
	horsepower	x 13.155	*x 13*
	kilowatts	x 9.809 5	*x 10*
	square feet of heating surface	x 10^	
	watts	x 9 809.5	*x 10 000*
horsepower, electric	watts	x 746^	*x 750*
horsepower, metric	foot-pounds (force) per second	x 542.48	*x 540*
	horsepower	x 0.986 32	*x 1*
	kilocalories (IT) per second	x 0.175 67	*x 0.18*
	kilowatts	x 0.735 50	*x 0.74*
	meter-kilograms (force) per second	x 75^	
	watts	x 735.50	*x 740*
horsepower-hour	foot-pounds (force)	x 1.98^ E+06	*x 2 E+06*
	horsepower-hours, metric	x 1.013 9	*x 1*
	joules	x 2.684 5 E+06	*x 2.7 E+06*
	kilocalories (IT)	x 641.19	*x 640*
	kilowatt-hours	x 0.745 70_	*x 0.75*
	liter-atmospheres, standard	x 26 494	*x 26 000*
	megajoules	x 2.684 5	*x 2.7*
	meter-kilograms (force)	x 2.737 4 E+05	*x 2.7 E+05*

TABLE 2
Measurement Conversions, Alphabetical

All measurement units are US, unless otherwise noted. All number denominations "billion" and higher are US, unless otherwise noted.

CONVERT	TO EQUIVALENT	BY PRECISELY	OR WITHIN ± 5.0 %
horsepower-hour, metric	foot-pounds (force)	x 1.952 9 E+06	*x 2 E+06*
	horsepower-hours	x 0.986 32	*x 1*
	joules	x 2.647 8 E+06	*x 2.6 E+06*
	kilocalories (IT)	x 632.42	*x 630*
	kilowatt-hours	x 0.735 50	*x 0.74*
	liter-atmospheres, standard	x 26 132	*x 26 000*
	meter-kilograms (force)	x 2.700 0 E+05	*x 2.7 E+05*
hour (customary)	hour, mean solar	x 1^	
hour, inequal (for astrology)	natural day or natural night (according to times of sunrise and sunset)	/12^ or x 0.083 333	*x 0.08*
hour, mean solar	days	/24^ or x 0.041 667	*x 0.04*
	hours, sidereal	x 1.002 7	*x 1*
	minutes	x 60^	
	seconds	x 3 600^	
hour, planetary	hour, inequal	x 1^	
hour, sidereal	days, sidereal	/24^ or 0.041 667	*x 0.04*
	mean solar seconds	x 3 590.2	*x 3 600*
	minutes, sidereal	x 60^	
	seconds, sidereal	x 3 600^	
human sensitivity to sound	See "range"		
humidity, molal	mols of water per mol of dry gas	x 1^	
humidity, specific	mass of water vapor per unit mass of moist air	x 1^	
hundredweight (Brit.)	pounds, avoirdupois	x 112^	*x 110*
	stones (Brit.)	x 8^	
hundredweight, gross	hundredweight, long	x 1^	
hundredweight, long	hundredweight, gross	x 1^	
	kilograms	x 50.802	*x 50*
	pounds	x 112^	*x 110*
hundredweight, long (Can.)	hundredweight, long (US)	x 1^	
hundredweight, long (US)	hundredweight (Brit.)	x 1^	
hundredweight, net	hundredweight, short	x 1^	
hundredweight, short	hundredweight, net	x 1^	
	kilograms	x 45.359	*x 45*
	ounces, avoirdupois	x 1 600^	
	pounds, avoirdupois	x 100^	
	tons, long	x 0.044 643	*x 0.045*
	tons, metric	x 0.045 359	*x 0.045*
	tons, short	/20^ or x 0.05^	
hundredweight, short (Can.)	hundredweight, short (US)	x 1^	
hypersonic	Mach above 5	x 1	
-I-			
impedance	1/admittance	x 1^	
inch	angstroms	x 2.54^E+08	*x 2.5 E+08*
	centimeters	x 2.54^	*x 2.5*
	feet	/12^ or x 0.083 333	*x 0.08*
	hands (for height of horses)	/4 or x 0.25^	
	meters	x 0.025 4^	*x 0.025*
	microns	x 2.54^E+04	*x 2.5 E+04*
	miles	/63 360^ or x 1.578 3 E-05	*x 1.6 E-05*

TABLE 2
Measurement Conversions, Alphabetical

All measurement units are US, unless otherwise noted. All number denominations "billion" and higher are US, unless otherwise noted.

CONVERT	TO EQUIVALENT	BY PRECISELY	OR WITHIN ± 5.0 %
	millimeters	x 25.4^	x 25
	mils (length)	x 1 000^	
	points (printer's)	x 72.281	x 72
	yards	/36^ or x 0.027 778	x 0.028
inch of mercury (@ 60 deg. F)	pascals	x 3 376.9	x 3 400
	pounds (force) per square inch	x 0.489 77	x 0.5
inch of mercury [13]	atmospheres, standard	x 0.033 421	x 0.033
	bars	x 0.033 864	x 0.034
	kilograms (force) per square centimeter	x 0.034 532	x 0.035
	pascals	x 3 386.4	x 3 400
inch of rain	cubic feet of water per acre	x 3 630^	x 3 600
	gallons per square yard	x 5.610 4	x 5.6
inch of water (@ 60 deg. F)	pascals	x 248.84	x 250
	pounds (force) per square inch	x 0.036 091	x 0.036
inch of water [13]	pascals	x 249.08	x 250
inch per second	meters per second	x 0.025 4^	x 0.025
inch per second per second	centimeters per second per second	x 2.54^	x 2.5
	meters per second per second	x 0.025 4^	/40
inch (Brit.)	inch (US)	x 1^	
inch (Can.)	inch (US)	x 1^	
inch, miner's	cubic feet per minute	x 1.5 (usually)	
inch, solid	cubic inch	x 1^	
inch, superficial	square inch	x 1^	
infinity	a number without end	x 1	
infrasound	See "range, infrasound"		
ips	instructions per second	x 1^	
-J-			
jeroboam (for wine, Brit.)	bottles (Brit.)	x 4^	
	liters	x 3.0	
joule	Btu (IT)	x 9.478 2 E-04	x 9.5 E-04
	Btu (@ 39 deg. F)	x 9.436 9 E-04	x 9.4 E-04
	Btu (@ 59 deg. F)	x 9.480 5 E-04	x 9.5 E-04
	Btu (@ 60 deg. F)	x 9.481 5 E-04	x 9.5 E-04
	Btu, mean	x 9.470 9 E-04	x 9.5 E-04
	Btu, thermochemical	x 9.484 5 E-04	x 9.5 E-04
	calories (IT)	x 0.238 85	x 0.24
	calories (@ 15 deg. C)	x 0.238 90	x 0.24
	calories (@ 20 deg. C)	x 0.239 13	x 0.24
	calories, mean	x 0.238 66	x 0.24
	calories, thermochemical	x 0.239 01	x 0.24
	centigrade heat units	x 5.265 6 E-04	x 5.3 E-04
	electronvolts	x 6.241 5 E+18	x 6 E+18
	foot-pounds (force)	x 0.737 56	x 0.74
	grams (equivalent mass)	x 1.112 7 E-14	x 10/9 E-14
	horsepower-hours	x 3.725 1 E-07	x 3.7 E-07
	horsepower-hours, metric	x 3.776 7 E-07	x 3.8 E-07
	kilocalories (IT)	x 2.388 5 E-04	x 2.4 E-04

[13] Unless otherwise noted, liquid-head conversions are based on: a pressure of one standard atmosphere; temperature for mercury = 0.0 deg. C = 32.0 deg. F; temperature for water = 4.0 deg. C = 39.2 deg. F.

TABLE 2
Measurement Conversions, Alphabetical

All measurement units are US, unless otherwise noted. All number denominations "billion" and higher are US, unless otherwise noted.

CONVERT	TO EQUIVALENT	BY PRECISELY	OR WITHIN ± 5.0 %
	kilocalories, mean	x 2.386 6 E-04	*x 2.4 E-04*
	kilocalories, thermochemical	x 2.390 1 E-04	*x 2.4 E-04*
	liter-atmospheres, standard	x 0.009 869 2	*x 0.01*
	meter-kilograms (force)	x 0.101 97	*x 0.1*
	meter-newton	x 1^	
	quads (energy)	x 9.479 E-19	*x 9.5 E-19*
	therms (EEC)	x 9.478 1 E-09	*x 9.5 E-09*
	therms (US)	x 9.480 4 E-09	*x 9.5 E-09*
	watt-second	x 1^	
joule per kilowatt-hour (power-plant heat rate)	Btu (IT) per kilowatt-hour	x 9.478 2 E-04	*x 9.5 E-04*
joule per second	watt	x 1^	
joule, absolute	joules, international	x 0.999 84	*x 1*
Julian century	days	x 36 525	*x 37 000*
Julian day calendar (for astronomy)	a calendar whose starting date is the year 4713 B.C.		
Julian day number	Julian-day-calendar sequential number for any specific day [18]	x 1	
-K-			
K	kilobyte	x 1^	
karat (for gemstones)	See "carat"		
karat (of gold) [2]	grams of gold per kilogram of alloy	x 125/3^ or x 41.667	*x 40*
	percent by weight of gold in alloy	x 25/6^ or x 4.166 7	*x 4*
kayser	wave number	x 1^	
kelvin (span)	deg. Celsius	x 1^	
	deg. Fahrenheit	x 9/5^ or x 1.8^	
	deg. Rankine	x 9/5^ or x 1.8^	
kelvin (temp.)	deg. Celsius	- 273.15	*- 270*
	deg. Fahrenheit	x 9/5^ - 459.67	*x 1.8 - 460*
	deg. Rankine	x 9/5^ or x 1.8^	
kilderkin (Brit.)	barrels, Imperial (Brit.)	/2 or x 0.5	
	gallons, Imperial (Brit.)	x 18	
kilo (short form)	kilogram	x 1^	
kilobaud	bauds	x 2^ to power 10 or x 1 024^	*x thousand*
kilobit	bits	x 2^ to power 10 or x 1 024^	*x thousand*
kilobyte	bytes	x 2^ to power 10 or x 1 024^	*x thousand*
kilocalorie	Btu (IT)	x 3.968 3	*x 4*
	calories	x 1 000^	
kilocalorie (IT)	Btu (IT)	x 3.968 3	*x 4*
	foot-pounds (force)	x 3 088.0	*x 3 000*
	horsepower-hours	x 0.001 559 6	*x 0.001 6*
	horsepower-hours, metric	x 0.001 581 2	*x 0.001 6*
	joules	x 4 186.8^	*x 4 000*
	kilowatt-hours	x 0.001 163^	*x 0.001 2*
	liter-atmospheres, standard	x 41.321	*x 40*
	meter-kilograms (force)	x 426.93	*x 430*
kilocalorie (IT) per minute	foot-pounds (force) per second	x 51.467	*x 50*
	horsepower	x 0.093 577	*x 0.09*
	kilowatts	x 0.069 78^	*x 0.07*

[2] 100-percent pure gold has 24 karats of gold.
[18] Julian-day-number example: January 1, 1960 is day number 2 346 934.

TABLE 2
Measurement Conversions, Alphabetical

All measurement units are US, unless otherwise noted. All number denominations "billion" and higher are US, unless otherwise noted.

CONVERT	TO EQUIVALENT	BY PRECISELY	OR WITHIN ± 5.0 %
	meter-kilograms (force) per second	x 7.115 6	*x 7*
	watts	x 69.78^	*x 70*
kilocalorie (IT) per second	foot-pounds (force) per second	x 3 088.0	*x 3 000*
	horsepower	x 5.614 6	*x 5.6*
	horsepower, metric	x 5.692 5	*x 5.7*
	kilowatts	x 4.186 8^	*x 4*
	meter-kilograms (force) per second	x 426.93	*x 430*
	watts	x 4 186.8^	*x 4 000*
kilocalorie (IT) (for foods and biological heat output) [5]	joules	x 4 186.8^	*x 4 000*
kilocalorie, (thermochemical)	joules	x 4 184^	*x 4 000*
kilocalorie (thermochemical) per minute	watts	x 1 046/15^ or x 69.733	*x 70*
kilocalorie (thermochemical) per second	foot-pounds (force) per second	x 3 086.0	*x 3 000*
	horsepower	x 5.610 8	*x 5.6*
	horsepower, metric	x 5.688 7	*x 5.7*
	kilowatts	x 4.184^	*x 4*
	meter-kilograms (force) per second	x 426.65	*x 430*
kilocalorie, mean (over range 0 to 100 deg. C)	joules	x 4 190.0	*x 4 000*
kilocycles per second	cycles per second	x 1 000^	
kilogram	carats	x 5 000^	
	drams, apothecary	x 257.21	*x 260*
	drams, avoirdupois	x 564.38	*x 560*
	grains	x 15 432	*x 15 000*
	grams	x 1 000^	
	hundredweights, long	x 0.019 684	*x 0.02*
	hundredweights, short	x 0.022 046	*x 0.022*
	milligrams	x 1^E+06	
	ounces, advoirdupois	x 35.274	*x 35*
	ounces, apothecary	x 32.151	*x 32*
	ounces, troy	x 32.151	*x 32*
	pennyweights	x 643.01	*x 640*
	pounds	x 2.204 6	*x 2.2*
	pounds, apothecary	x 2.679 2	*x 2.7*
	pounds, avoirdupois	x 2.204 6	*x 2.2*
	pounds, troy	x 2.679 2	*x 2.7*
	scruples, apothecary	x 771.62	*x 800*
	tons, long	x 9.842 1 E-04	*x 0.001*
	tons, metric	/1 000^ or x 0.001^	
	tons, short	x 0.001 102 3	*x 0.001 1*
kilogram per cubic meter	grains per gallon	x 58.418	*x 60*
	gram per liter	x 1^	
	grams per milliliter	/1 000^ or x 0.001^	
	ounces per cubic inch	x 5.780 4 E-04	*x 6 E-04*

5 The energy value of food and drink is customarily stated in "calories", but the technically correct measuring unit is "kilocalories", which is sometimes called "large calories". (1 kilcalorie = 1 000 calories) A person on a reducing diet may take 1 100 "calories" per day, but, in truth, he is taking 1 100 kilocalories, or 1.1 million (1 100 000) calories, or 1.1 megacalories.

TABLE 2
Measurement Conversions, Alphabetical

All measurement units are US, unless otherwise noted. All number denominations "billion" and higher are US, unless otherwise noted.

CONVERT	TO EQUIVALENT	BY PRECISELY	OR WITHIN ± 5.0 %
	ounces per gallon	x 0.133 53	*x 0.13*
	ounces per gallon, Imperial (Brit.)	x 0.160 36	*x 0.16*
	pounds per cubic foot	x 0.06 242 8	*x 0.06*
kilogram per second	pounds per hour	x 7 936.6	*x 8 000*
	pounds per minute	x 132.28	*x 130*
	pounds per second	x 2.204 6	*x 2.2*
	tons, short per hour	x 3.968 3	*x 4*
kilogram (force)	newtons	x 9.806 7	*x 10*
kilogram (force) per meter	pounds (force) per foot	x 0.671 97	*x 0.7*
kilogram (force) per square centimeter	atmosphere, technical	x 1^	
	pascals	x 98 067	*x 1 E+05*
	pounds (force) per square inch	x 14.223	*x 100/7 or x 14*
kilogram (force) per square meter	pascals	x 9.806 7	*x 10*
	pounds (force) per square inch	x 0.001 422 3	*/700 or x 0.001 4*
kilogram (force) per square millimeter	pascals	x 9.806 7 E+06	*x 10 E+06*
kilogram (force)-meter	newton-meters	x 9.806 7	*x 10*
kilogram-calorie	kilocalorie	x 1^	
kilogram-square meter	gram-square centimeters	x 1^E+07	
	pound-square feet	x 23.730	*x 24*
	pound-square inches	x 3 417.2	*x 3 400*
	slug-square feet	x 0.737 56	*x 0.74*
kilojoule per kilowatt-hour (power-plant heat rate)	Btu (IT) per kilowatt-hour	x 0.947 82	*x 0.95*
kilometer	leagues, land	x 0.207 12	*x 0.2*
	marathons	x 0.023 699	*x 0.024*
	miles	x 0.621 37	*x 5/8 or x 0.6*
	miles (intl.)	x 0.621 37	*x 5/8 or x 0.6*
	miles, nautical (intl.)	x 0.539 96	*x 0.54 or x 5/9*
	miles, nautical (US)	x 0.539 96	*x 0.54 or x 5/9*
	miles, statute (intl.)	x 0.621 37	*x 5/8 or x 0.6*
	miles, statute (US)	x 0.621 37	*x 5/8 or x 0.6*
kilometer per hour	centimeters per second	x 27.778	*x 28*
	feet per minute	x 54.681	*x 55*
	feet per second	x 0.911 34	*x 0.9*
	knots (intl.)	x 250/463^ or x 0.539 957	*x 0.54*
	meter per second	x 5/18^ or x 0.277 78	*x 0.28*
	meters per minute	x 50/3^ or x 16.667	*x 17*
	miles per hour	x 0.621 37	*x 0.6*
kilometer per hour per second	centimeters per second per second	x 250/9^ or x 27.778	*x 28*
	feet per second per second	x 0.911 34	*x 0.9*
	meters per second per second	x 5/18^ or x 0.277 78	*x 0.28*
kilometer per liter (vehicle fuel economy)	miles per gallon	x 2.352 1	*x 2.4*
kilonewton	tons, short (force)	x 0.112 40	*x 0.11*
kilopascal	bars	/100^ or x 0.01^	
	pounds (force) per square inch	x 0.145 04	*x 0.15*

TABLE 2
Measurement Conversions, Alphabetical

All measurement units are US, unless otherwise noted. All number denominations "billion" and higher are US, unless otherwise noted.

CONVERT	TO EQUIVALENT	BY PRECISELY	OR WITHIN ± 5.0 %
kilopond	kilogram (force, @ standard gravity) [29][8]	x 1^	
kilowatt	Btu per minute	x 56.869	*x 57*
	foot-pounds (force) per minute	x 44 254	*x 44 000*
	foot-pounds (force) per second	x 737.56	*x 740*
	horsepower	x 1.341 0	*x 1.3*
	horsepower, metric	x 1.359 6	*x 1.4*
	kilocalories (IT) per minute	x 14.331	*x 14*
	kilocalories (IT) per second	x 0.238 85	*x 0.24*
	meter-kilograms (force) per second	x 101.97	*x 100*
kilowatt-hour	Btu (IT)	x 3 412.1	*x 3 400*
	calories (IT)	x 8.598 5 E+05	*x 9 E+05*
	ergs	x 3.6^E+13	
	foot-pounds (force)	x 2.655 2 E+06	*x 2.7 E+06*
	horsepower-hours	x 1.341 0	*x 1.3*
	horsepower-hours, metric	x 1.359 6	*x 1.4*
	joules	x 3.6^E+06	
	kilocalories (IT)	x 859.85	*x 900*
	liter-atmospheres, standard	x 35 529	*x 36 000*
	megajoules	x 3.6^	
	meter-kilograms (force)	x 3.671 0 E+05	*x 3.7 E+05*
kip per square inch	kilopascals	x 6 894.8	*x 7 000*
	pounds (force) per square inch	x 1 000^	
kip (kilopound, force)	newtons	x 4 448.2	*x 4 400*
	pounds (force)	x 1 000^	
knot	miles, statute per hour	x 1.150 8	*x 1.2*
knot (Brit.)	knot (intl.)	x 1^	
knot (intl.)	feet per hour	x 6 945/1.143^ or x 6 076.1	*x 6 000*
	feet per second	x 1.687 8	*x 1.7*
	kilometers per hour	x 1.852^	*x 1.9*
	meters per second	x 463/900^ or x 0.514 44	*x 0.5*
	miles per hour (intl.)	x 1.150 8	*x 1.2*
	mile, nautical per hour (intl.)	x 1^	
	yards per hour	x 2 025.4	*x 2 000*
knot (US)	knot (intl.)	x 1^	
knot (US) [21]	mile, nautical (intl.) per hour	x 1^	
kondratieff cycle (for the business cycle)	years	x 50 to x 60	
k-value	thermal conductivity	x 1^	
-L-			
labor (Texas)	acres	x 177.1	*x 180*
lambda	cubic millimeter	x 1^	
lambert	candelas per square centimeter	/pi or x 0.318 31	*x 0.32*
	candelas per square foot	x 295.72	*x 300*
	candelas per square inch	x 2.053 6	*x 2*
	candelas per square meter	x 10 000/pi or x 3 183.1	*x 3 200*
	lumen per square centimeter	x 1^	
langley (solar radiation)	calorie per square centimeter	x 1^	

[8] Standard conditions for the acceleration of gravity are: 9.806 65 meters per second per second = 32.174 0 feet per second per second, at sea level and latitude 45 degrees. At other locations, the acceleration may differ within a span of more than 0.5 percent of the standard value.

[21] A ship's log line for measuring speed is marked in segments that are knotted every 47 feet, 3 inches, which equals 14.402 meters. The ship's speed in nautical miles per hour, or knots, equals almost exactly the number of line segments or knots that are counted in 28 seconds as the line is unreeled.

TABLE 2
Measurement Conversions, Alphabetical

All measurement units are US, unless otherwise noted. All number denominations "billion" and higher are US, unless otherwise noted.

CONVERT	TO EQUIVALENT	BY PRECISELY	OR WITHIN ± 5.0 %
	joules per square meter	x 41 840^	*x 40 000*
latitude	angular distance, in degrees, from a given point to a specified circle or reference plane [1]	x 1^	
league (Texas)	acres	x 4 428.4	*x 4 400*
league, land	kilometers	x 4.828 0	*x 5*
	miles	x 3^	
league, marine	league, nautical	x 1^	
league, nautical	miles, nautical	x 3^	
league, sea	league, nautical	x 1^	
light-year	astronomical units	x 63 240	*x 63 000*
	kilometers	x 9.460 6 E+12	*x 9.5 E+12*
	meters	x 9.460 6 E+15	*x 9.5 E+15*
	miles	x 5.878 5 E+12	*x 6 E+12*
	parsecs	x 0.306 60	*x 0.3*
ligne (for buttons)	line (for buttons)	x 1^	
line	maxwell	x 1^	
line per square centimeter	gauss	x 1^	
line per square inch	gauss	x 0.155 00	*x 0.16*
	webers per square centimeter	x 1.550 0 E-09	*x 1.6 E-09*
	webers per square inch	x 1^E-08	
	webers per square meter	x 1.550 0 E-05	*x 1.6 E-05*
line (for buttons)	inches	/40^	
line (for fisherman's setline)	fathoms	x 50	
line (for whaling harpoons)	fathoms	x 150 (approx.)	
line (length, obsolete)	inches	/12^	
link (Can.)	link, surveyor's (US)	x 1^	
link, engineer's	inches	x 12^	
link, surveyor's	centimeters	x 20.117	*x 20*
	chains, surveyor's	/100^ or x 0.01^	
	feet	x 0.66^	
	inches	x 7.92^	*x 8*
	meters	x 0.201 17	*x 0.2*
	miles	/8 000^ or x 1.25^E-04	*x 1.3 E-04*
	rods (length)	/25^ or x 0.04^	
liter	bushels, struck measure	x 0.028 378	*x 0.28*
	cubic decimeter	x 1^	
	cubic feet	x 0.035 315	*x 0.035*
	cubic inches	x 61.024	*x 60*
	cubic meters	/1 000^ or x 0.001^	
	cubic yards	x 0.001 308 0	*x 0.001 3*
	drams, fluid	x 270.51	*x 270*
	gallons	x 0.264 17	*x 0.26 or x 3/11*
	gallons, Imperial (Brit.)	x 0.219 97	*x 0.22 or x 2/9*
	gills	x 8.453 5	*x 8.5*
	milliliters	x 1 000^	
	minims	x 16 231	*x 16 000*
	ounces, fluid	x 33.814	*x 34*
	pecks	x 0.113 51	*x 0.11*

1 "Latitude" example: A latitude of the Earth is an angular distance north or south along a meridian, measured in the range of 0 to 90 degrees from the equator.

TABLE 2
Measurement Conversions, Alphabetical

All measurement units are US, unless otherwise noted. All number denominations "billion" and higher are US, unless otherwise noted.

CONVERT	TO EQUIVALENT	BY PRECISELY	OR WITHIN ± 5.0 %
	pints, dry	x 1.816 2	*x 1.8*
	pints, fluid	x 2.113 4	*x 2.1*
	quarts, dry	x 0.908 08	*x 0.9*
	quarts, fluid	x 1.056 7	*x 1.1*
liter per minute	cubic feet per second	x 5.885 8 E-04	*x 6 E-04*
	gallons per minute	x 0.264 17	*x 0.26*
	gallons per second	x 0.004 402 9	*x 0.004 4*
	gallons, Imperial per second	x 0.003 666 2	*x 0.003 7*
liter per second	barrels (42 gallons) per day	x 543.44	*x 540*
	barrels (42 gallons) per hour	x 22.643	*x 23*
	cubic feet per hour	x 127.13	*x 130*
	cubic feet per minute	x 2.118 9	*x 2.1*
	cubic feet per second	x 0.035 315	*x 0.035*
	gallons per hour	x 951.02	*x 950*
	gallons per minute	x 15.850	*x 16*
liter-atmosphere, standard	foot-pounds (force)	x 74.733	*x 75*
	horsepower-hours	x 3.774 4 E-05	*x 3.8 E-05*
	horsepower-hours, metric	x 3.826 8 E-05	*x 3.8 E-05*
	joules	x 101.32	*x 100*
	kilocalories (IT)	x 0.024 201	*x 0.024*
	kilograms (force)	x 10.332	*x 10*
	kilowatt-hours	x 2.814 6 E-05	*x 2.8 E-05*
litre	liter (US)	x 1^	
ln N (N, a number)	log N	x 2.302 59	*x 2.3*
	$log_e N$	x 1^	
	natural logarithm of N	x 1^	
log N (N, a number)	common logarithm of N	x 1^	
	ln N	x 0.434 294	*x 0.43*
	$log_{10} N$	x 1^	
logarithm, Briggs	common logarithm	x 1^	
logarithm, common	logarithm to base *10*	x 1^	
logarithm, denary	common logarithm	x 1^	
logarithm, hyperbolic	natural logarithm	x 1^	
logarithm, Napierian	natural logarithm	x 1^	
logarithm, natural	logarithm to base *e*	x 1^	
$log_{10} N$ (N, a number)	common logarithm	x 1^	
	log N	x 1^	
$log_e N$ (N, a number)	ln N	x 1^	
longitude	angular distance, in degrees or time, from a point east or west to the prime meridian in Greenwich, England	x 1^	
lumber, 2 x 4 nominal	1-5/8 x 3-5/8 inches (dressed)	x 1	
lumen	spherical candle power	/4 pi or x 0.079 577	*x 0.08*
lumen per square foot	foot-candle	x 1^	
	lumens per square meter	x 10.764	*x 11*
lumen per square meter	lumens per square foot	x 0.092 903	*x 0.09*
	lux	x 1^	
	phots	/10 000^ or x 1^E-04	
luminous flux	lumen	x 1^	
lunation	lunar month	x 1^	
lux	lumen per square meter	x 1^	
	foot-candles	x 0.092 903	*x 0.09*

TABLE 2
Measurement Conversions, Alphabetical

All measurement units are US, unless otherwise noted. All number denominations "billion" and higher are US, unless otherwise noted.

CONVERT	TO EQUIVALENT	BY PRECISELY	OR WITHIN ± 5.0 %
-M-			
M	Mach number	x 1^	
Mach	Mach number	x 1^	
Mach 1 to 5	supersonic speed	x 1	
Mach approximately 1	transonic speed	x 1	
Mach greater than 5	hypersonic speed	x 1	
Mach less than 1	subsonic speed	x 1	
Mach number	ratio of object speed to sound speed in a compressible fluid	x 1^	
magnetic field strength	ampere per meter	x 1^	
magnification	ratio of optical-image size to object size	x 1^	
magnitude decrease of one (for relative brightness of celestial body)	apparent brightness increase	x 100^ to power 0.2 or x 2.511 9	*x 2.5*
magnum (for wine, Brit.)	bottles (Brit.)	x 2^	
	liters	x 1.5	
marathon	26 miles, 385 yards	x 1^	
	kilometers	x 42.195	*x 42*
	miles	x 26.219	*x 26*
mass	weight (force)	/g (see LINEAR ACCELERATION, "g")	
mass unit	atomic mass unit	x 1^	
mass, molar	kilogram per mole	x 1^	
maxwell	line	x 1^	
	webers	x 1^E-08	
MB	megabyte	x 1^	
megabyte	bytes	x 2^ to power 20 or x 1.048 6 E+06	*x million*
megacycles per second	cycles per second	x 1^E+06	
megadyne per square centimeter	bar	x 1^	
megaflops	million floating-point operations per second	x 1^	
megajoule	horsepower-hours	x 0.373 13	*x 0.37*
	kilowatt-hours	x 5/18^ or x 0.277 78	*x 0.28*
megaton (explosive force)	tons of TNT	x 1^E+06	
megawatt-hour per kilogram	joules per kilogram	x 3.6^E+09	
Mercalli scale (for earthquakes)	See APPENDIX, "earthquakes"		
meter	angstroms	x 1^E+10	
	centimeters	x 100^	
	chains, engineer's	x 0.032 808	*x 0.033*
	chains, surveyor's	x 0.049 710	*x 0.05*
	fathoms	x 0.546 81	*x 0.55*
	feet	x 1 250/381^ or x 3.280 8	*x 3.3*
	furlongs	x 0.004 971 0	*x 0.005*
	inches	x 39.370	*x 40*
	links, engineer's	x 3.280 8	*x 3.3*
	links, surveyor's	x 4.971 0	*x 5*
	microinches	x 3.937 0 E+07	*x 4 E+07*
	micromicrons	x 1^E+12	
	miles	x 6.213 7 E-04	*x 6 E-04*
	miles, nautical (intl.)	x 5.399 6 E-04	*x 5.4 E-04*

TABLE 2
Measurement Conversions, Alphabetical

All measurement units are US, unless otherwise noted. All number denominations "billion" and higher are US, unless otherwise noted.

CONVERT	TO EQUIVALENT	BY PRECISELY	OR WITHIN ± 5.0 %
	miles, nautical (US)	x 5.399 6 E-04	*x 5.4 E-04*
	miles, statute (US)	x 6.213 7 E-04	*x 6 E-04*
	mils (length)	x 39 370	*x 40 000*
	rods (length)	x 0.198 84	*x 0.2*
	yards	x 1.093 6	*x 1.1*
meter per minute	centimeters per second	x 5/3^ or x 1.666 7	*x 1.7*
	feet per minute	x 3.280 8	*x 3.3*
	feet per second	x 0.054 681	*x 0.055*
	kilometers per hour	x 0.06^	
	kilometers per minute	/1 000^ or x 0.001^	
	miles per hour	0.037 282	*x 0.037*
	miles per minute	x 6.213 7 E-04	*x 6 E-04*
meter per second	feet per minute	x 196.85	*x 200*
	feet per second	x 3.280 8	*x 3.3*
	kilometers per hour	x 3.6^	
	miles per hour	x 2.236 9	*x 2.2*
meter per second per second	feet per second per second	x 3.280 8	*x 3.3*
	kilometers per hour per second	x 3.6^	
	miles per hour per second	x 2.236 9	*x 2.2*
meters (focal length of lens)	1/diopters (of optical power)	x 1^	
meter-kilogram (force)	Btu (IT)	x 0.009 294 9	*x 0.009*
	calories (IT)	x 2.342 2	*x 2.3*
	centigrade heat units	x 0.005 163 8	*x 0.005*
	foot-pounds (force)	x 7.233 0	*x 7*
	horsepower-hours	x 3.653 0 E-06	*x 3.7 E-06*
	horsepower-hours, metric	x 3.703 7 E-06	*x 3.7 E-06*
	joules	x 9.806 6	*x 10*
	kilocalories (IT)	x 0.002 342 3	*x 0.002 3*
	kilowatt-hours	x 2.724 1 E-06	*x 2.7 E-06*
	liter-atmospheres, standard	x 0.096 784	*x 0.1*
meter-kilogram (force) per second	foot-pounds (force) per second	x 7.233 0	*x 7*
	horsepower	x 0.013 151	*x 0.013*
	horsepower, metric	/75^ or x 0.013 333	*x 0.013*
	kilocalories (IT) per second	x 0.002 342 3	*x 0.002 3*
	kilowatts	x 0.009 806 6	*x 0.001*
meter-newton	joule	x 1^	
methuselah (for wine, Brit.)	bottles (Brit.)	x 8^	
	liters	x 6.0	
metonic cycle	See "cycle, metonic"		
metre	meter (US)	x 1^	
metric horsepower-hour	See "horsepower-hour, metric"		
metric prefixes	See Table 1		
mev	million electron-volts	x1^	
mho	1/ohm	x 1^	
	siemens	x 1^	
microbar	pascals	/10^ or x 0.1^	
microgram	grains	x 1.543 2 E-05	*x 1.5 E-05*
	grams	x 1^E-06	
microhm per centimeter cube	microhm-centimeter	x 1^	

TABLE 2
Measurement Conversions, Alphabetical

All measurement units are US, unless otherwise noted. All number denominations "billion" and higher are US, unless otherwise noted.

CONVERT	TO EQUIVALENT	BY PRECISELY	OR WITHIN ± 5.0 %
microhm per inch cube	microhm-inch	x 1^	
microhm-centimeter	microhm-inches	x 50/127^ or x 0.393 70	*x 0.4*
	ohms (mil, foot)	x 6.015 3	*x 6*
	ohm-meters	x 1^E-08	
microhm-inch	microhm-centimeters	x 2.54^	*x 2.5*
	ohms (mil, foot)	x 15.279	*x 15*
	ohm-meters	x 2.54^E-08	*x 2.5 E-08*
microinch	inches	x 1^E-06	
	meters	x 2.54^E-08	*x 2.5 E-08*
	millimeters	x 2.54^E-05	*x 2.5 E-05*
micrometer	millimeters	/1 000^ or x 0.001^	
micromicron (obsolete)	meters	x 1^E-12	
micron	inches	x 3.937 0 E-05	*x 4 E-05*
	meters	x 1^E-06	
micron of mercury (@ 0 deg. C)	millitorrs	/1 000 or x 0.001	
mil (angle, for artillery)	degrees (angle)	x 9/160^ or x 0.056 25^	*x 0.056*
	minutes (angle)	x 27/8^ or x 3.375^	*x 3.4*
	radians	x pi/3 200 or x 9.817 5 E-04	*x 0.001*
	revolutions	/6 400^ or x 1.562 5^E-04	*x 1.6*
	seconds (angle)	x 202.5^	*x 200*
mil (length)	inches	/1 000^ or x 0.001^	
	meters	x 2.54^E-05	*x 2.5 E-05*
	millimeters	x 0.025 4^	*x 0.025*
Milankovitch Earth-axis-tilt cycle	years	x 41 000 (approx.)	
Milankovitch Earth-equinoctial-precession cycle	years	x 28 500 (approx.)	
Milankovitch Earth-orbit-eccentricity cycle	years	x 100 000 (approx.)	
mile	centimeters	x 1.609 3 E+05	*x 1.6 E+05*
	chains, surveyor's	x 80^	
	feet	x 5 280^	*x 5 300*
	furlongs	x 8^	
	inches	x 63 360^	*x 63 000*
	kilometers	x 1.609 3	*x 1.6*
	leagues, land	/3^ or x 0.333 33	*x 0.33*
	links, engineer's	x 5 280^	*x 5 300*
	links, surveyor's	x 8 000^	
	meters	x 1 609.3	*x 1 600*
	mile, nautical	x 0.868 98	*x 13/15 or x 7/8*
	mile, statute	x 1^	
	rods (length)	x 320^	
	yards	x 1 760^	*x 1 800*
mile of line	distance between two points connected by rail line	x 1^	
mile of road	mile of line	x 1^	
mile per gallon (vehicle fuel economy)	kilometers per liter	x 0.425 14	*x 0.43*
mile per hour	centimeters per second	x 44.704^	*x 45*
	feet per minute	x 88^	*x 90*

TABLE 2
Measurement Conversions, Alphabetical

All measurement units are US, unless otherwise noted. All number denominations "billion" and higher are US, unless otherwise noted.

CONVERT	TO EQUIVALENT	BY PRECISELY	OR WITHIN ± 5.0 %
	feet per second	x 22/15^ or x 1.466 7	*x 1.5*
	kilometers per hour	x 1.609 3	*x 1.6*
	knots	x 0.868 98	*x 0.87*
	meters per minute	x 26.822	*x 27*
	meters per second	x 0.447 04^	*x 0.45*
	miles, nautical per hour	x 0.868 98	*x 0.87*
mile per hour per second	feet per second per second	x 22/15^ or x 1.466 7	*x 1.5*
	kilometers per hour per second	x 1.609 3	*x 1.6*
	meters per second per second	x 0.447 04	*x 0.45*
mile per minute	centimeters per second	x 2 682.2	*x 2 700*
	feet per second	x 88^	*x 90*
	kilometers per minute	x 1.609 3	*x 1.6*
	meters per second	x 26.822	*x 27*
	miles per hour	x 60^	
mile per second	meters per second	x 1 609.3	*x 1 600*
mile (Brit.)	feet	x 5 280^	*x 5 300*
	feet	x 5 000^ (obsolete)	
	kilometers	x 1.609 3	*x 1.6*
	meters	x 1 609.3	*x 1 600*
	mile (US)	x 1^	
mile (Can.)	mile, statute (US)	x 1^	
mile (intl.)	feet (intl.)	x 5 280^	*x 5 300*
	kilometers	x 1.609 3	*x 1.6*
	meters	x 1 609.3	*x 1 600*
	mile (US)	x 1	
	mile, statute (US)	x 1	
mile (intl.) per hour	mile per hour	x 1	
mile (US)	leagues (British)	/3 or x 0.3	
mile, Admiralty (Brit.)	mile, nautical (Brit.)	x 1^	
mile, air	feet (1946 - 1954)	x 6 080.2	*x 6 100*
	feet (before 1946)	x 5 280^	*x 5 300*
	feet (since 1954)	x 6 076.1	*x 6 000*
	meters (since 1954)	x 1 852^	*x 1 900*
mile, air (US)	mile, nautical (Brit., 1946 - 1954)	x 1^	
	mile, nautical (intl., since 1954)	x 1^	
	mile, statute (US, before 1946)	x 1^	
mile, geographic	mile, nautical	x 1^	
mile, nautical	feet	x 6 076.1 (since 1959)	*x 6 000*
	feet	x 6 080.2 (obsolete)	*x 6 100*
	kilometers	x 1.852^ (since 1959)	*x 1.9*
	leagues, nautical	/3^ or x 0.333 33	*x 0.33*
	meters	x 1 852^ (since 1959)	*x 1 900*
mile, nautical per hour	miles per hour	x 1.150 8	*x 1.2*
mile, nautical (Brit.)	feet	x 6 076.1	*x 6 000*
	feet	x 6 080 (obsolete)	*x 6 000*
	kilometers	x 1.852^	*x 1.9*
	meters	x 1 853.2 (obsolete)	*x 1 900*
	meters	x 1 852^	*x 1 900*
mile, nautical (Can.)	meters	x 1 852^	
mile, nautical (intl.)	feet	x 6 076.1	*x 6 000*
	kilometers	x 1.852^	*x 1.9*

TABLE 2
Measurement Conversions, Alphabetical

All measurement units are US, unless otherwise noted. All number denominations "billion" and higher are US, unless otherwise noted.

CONVERT	TO EQUIVALENT	BY PRECISELY	OR WITHIN ± 5.0 %
	meters	x 1 852^	*x 1 900*
	mile, nautical (US)	x 1^	
	mile, statute (US)	x 1.150 8	*x 1.2*
mile, sea	mile, nautical	x 1^	
mile, statute	feet, survey	x 5 280	*x 5 300*
	meters	x 1 609.3	*x 1 600*
	mile	x 1^	
	mile, survey	x 1	
mile, statute per hour	mile per hour	x 1^	
millenium	years	x 1 000^	
milliard (Brit.)	billion (US)	x 1^	
millibar	pascals	x 100^	
	pounds (force) per square inch	x 0.014 504	*x 0.015*
millier	kilograms	x 1 000^	
	ton, metric	x 1^	
milligram	carats	/200^ or x 0.005^	
	drams, apothecary	x 2.572 1 E-04	*x 2.5 E-04*
	drams, avoirdupois	x 5.643 8 E-04	*x 5.6 E-04*
	grains	x 0.015 432	*x 0.015*
	grams	/1 000^ or x 0.001^	
	kilograms	/1^E+06 or x 1^E-06	
	ounces, apothecary	x 3.215 1 E-05	*x 3.2 E-05*
	ounces, avoirdupois	x 3.527 4 E-05	*x 3.5 E-05*
	ounces, troy	x 3.215 1 E-05	*x 3.2 E-05*
	pennyweights	x 6.430 1 E-04	*x 6.4 E-04*
	points (jeweler's)	/2^ or x 0.5^	
	pounds, apothecary	x 2.679 2 E-06	*x 2.7 E-06*
	pounds, avoirdupois	x 2.204 6 E-06	*x 2.2 E-06*
	pounds, troy	x 2.679 2 E-06	*x 2.7 E-06*
	scruples, apothecary	x 7.716 2 E-04	*x 8 E-04*
milligram per liter (of water @ 4 deg. C)	parts per million (by weight)	x 1^	
milliliter	cubic centimeter	x 1^	
	cubic feet	x 3.531 5 E-05	*x 3.5 E-05*
	cubic inches	x 0.061 024	*x 0.06*
	cubic meters	x 1^E-06	
	cubic yards	x 1.308 0 E-06	*x 1.3 E-06*
	cups, measuring	x 0.004 226 8	*x 0.004 2*
	drams, fluid	x 0.270 51	*x 0.27*
	gallons	x 2.641 7 E-04	*x 2.6 E-04*
	gills	x 0.008 453 5	*x 0.008 5*
	liters	/1 000^ or x 0.001^	
	minims	x 16.231	*x 16*
	ounces, fluid	x 0.033 814	*x 0.034*
	pints, dry	x 0.001 816 2	*x 0.001 8*
	pints, fluid	x 0.002 113 4	*x 0.002 1*
	quarts, fluid	x 0.001 056 7	*x 0.001 1*
millimeter	inches	x 0.039 370	*x 0.4*
	microinches	x 39 370	*x 40 000*
	mils (length)	x 39.370	*x 40*
millimeter of mercury (intl.)	torr	x 1.000 0	*x 1*

TABLE 2
Measurement Conversions, Alphabetical

All measurement units are US, unless otherwise noted. All number denominations "billion" and higher are US, unless otherwise noted.

CONVERT	TO EQUIVALENT	BY PRECISELY	OR WITHIN ± 5.0 %
millimeter of mercury[13]	atmospheres, standard	x 0.001 315 8	*x 0.001 3*
	pascals	x 133.32	*x 130*
	torr	x 1.000 0	*x 1*
million gallons per day	cubic feet per second	x 1.547 2	*x 1.5*
million (Brit.)	million (US)	x 1^	
million (US)	1^E+06	x 1^	
millisecond	second	/1 000^ or x 0.001^	
mil-foot	one circular mil by one foot	x 1^	
miner's inch	See "inch, miner's"		
minim	cubic feet	x 2.175 8 E-06	*x 2.2 E-06*
	cubic inches	x 0.003 759 8	*x 0.003 8*
	drams, fluid	/60^ or x 0.016 667	
	drop, fluid (approx.)	x 1	
	gallons	/61 440^ or x 1.627 6 E-05	*x 1.6 E-05*
	gills	/1 920^ or x 5.208 3 E-04	*x 5 E-04*
	liters	x 6.161 2 E-05	*x 6 E-05*
	milliliters	x 0.061 612	*x 0.06*
	ounces, fluid	/480^ or x 0.002 083 3	*x 0.002*
	pints, fluid	/7 680^ or x 1.30 21 E-04	*x 1.3 E-04*
	quarts, fluid	/15 360^ or x 6.510 4 E-05	*x 6.5 E-05*
minim, apothecary	minim	x 1^	
minim, Imperial apothecary (Brit.)	minim, Imperial (Brit.)	x 1^	
minim, Imperial (Brit.)	drachms, Imperial fluid (Brit.)	/60^ or x 0.016 667	*x 0.017*
	scruples, Imperial fluid (Brit.)	/20^ or x 0.05^	
minute (angle)	degrees (angle)	/60^ or x 0.016 667	*x 0.017*
	radians	x pi/10 800 or x 2.908 9 E-04	*x 3 E-04*
	revolutions	/21 600^ or x 4.629 6 E-05	*x 4.6 E-05*
	seconds (angle)	x 60^	
minute (customary)	minute, mean solar	x 1^	
minute, mean solar	hours, mean solar	/60^ or x 0.016 667	*x 0.017*
	hours, sidereal	x 0.016 712	*x 0.017*
	minutes, sidereal	x 1.002 7	*x 1*
	seconds	x 60^	
	seconds, sidereal	x 60.164	*x 60*
minute, sidereal	hours, sidereal	/60^ or x 0.016 667	*x 0.017*
	minutes, mean solar	x 0.997 27	*x 1*
	seconds, mean solar	x 59.836	*x 60*
	seconds, sidereal	x 60^	
mips	milion instructions per second	x 1^	
Modified Mercalli intensity scale	See APPENDIX, "earthquakes"		
mol	mole	x 1^	
molal solution	See "concentration, molal"		
molality	molal solution	x 1^	
molar energy	joule per mole	x 1^	
molar entropy	joule per mole-kelvin	x 1^	
molar heat capacity	joule per mole-kelvin	x 1^	
molar solution	See "concentration, molar"		
molarity	molar solution	x 1^	

[13] Unless otherwise noted, liquid-head conversions are based on: a pressure of one standard atmosphere; temperature for mercury = 0.0 deg. C = 32.0 deg. F; temperature for water = 4.0 deg. C = 39.2 deg. F.

TABLE 2
Measurement Conversions, Alphabetical

All measurement units are US, unless otherwise noted. All number denominations "billion" and higher are US, unless otherwise noted.

CONVERT	TO EQUIVALENT	BY PRECISELY	OR WITHIN ± 5.0 %
mole fraction	moles of solute per mole of solution	x 1^	
mole percent	mole fraction	/100^ or x 0.01^	
	moles of solute per 100 moles of solution	x 1^	
mole (mass)	chemical mass, e.g.,grams or pounds, equal in number to molecular weight	x 1^	
mole (quantity)	elementary-particle type as specified	x 6.022 5 E+23	*x 6 E+23*
molecular mass, relative	molecular weight	x 1^	
molecular weight	sum of atomic weights of all the atoms in a molecule	x 1^	
month (customary)	days, calendar	x 28^, 29^, 30^, or 31^	
	month, mean solar	x 1^	
month, anomalistic	days (customary)	x 27.555	*x 28*
month, lunar	29 days 12 hr. 44 min. 2.8 sec. mean solar time	x 1	
	days, mean solar	x 29.531	*x 30*
	hours, mean solar	x 708.73	*x 700*
month, mean calendar	days	x 30.417	
month, nodical	27 days 5 hr. 5 min. 35.8 sec. mean solar time	x 1	
	mean solar days	x 27.212	*x 27*
month, sidereal	27 days 7 hr. 43 min. 11.5 sec. mean solar time	x 1	
	mean solar days	x 27.322	*x 27*
month, synodic	lunar month	x 1^	
month, tropical	27 days 7 hr. 43 min. 4.7 sec. mean solar time	x 1	
	days	x 27.322	*x 27*
	tropical years	/12^ or x 0.083 333	*x 0.08*
moon	lunar month	x 1^	
mpg	miles per gallon	x 1^	
myriagram (obsolete)	grams	x 10 000^	
myriameter (obsolete)	meters	x 10 000^	
M-flops	megaflops	x 1^	
-N-			
nail (for cloth)	inches	x 2.25^	*x 2.3*
	yards	/16^ or x 0.062 5^	*x 0.06*
nanometer	microns	/1 000^ or x 0.001^	
nanosecond	seconds	x 1^E-09	
naught	zero	x 1^	
nebuchadnezzar (for wine, Brit.)	bottles (Brit.)	x 20^	
	liters	x 15	
neper (for power)	decibels	x 8.685 9	*x 200/23 or x 9*
	ratio of two power levels	x e (= 2.718 3 to power {no. of nepers x 2}	
neutron per kilobarn	neutrons per square meter	x 1^E+25	
neutron rest mass	atomic mass units	x 1.008 7	*x 1*
newton	dynes	x 1^E+05	
	kilograms (force)	x 0.101 97	*x 0.1*
	kiloponds	x 0.101 97	*x 0.1*
	kips	x 2.248 1 E-04	*x 2.2 E-04*
	ounces (force)	x 3.596 9	*x 3.6*
	poundals	x 7.233 0	*x 7*
	pounds (force)	x 0.224 81	*x 0.22*
	tons, short (force)	x 1.124 0 E-04	*x 1.1 E-04*

TABLE 2
Measurement Conversions, Alphabetical

All measurement units are US, unless otherwise noted. All number denominations "billion" and higher are US, unless otherwise noted.

CONVERT	TO EQUIVALENT	BY PRECISELY	OR WITHIN ± 5.0 %
newton per meter	pounds (force) per foot	x 0.068 522	*x 0.07*
	pounds (force) per inch	x 0.571 01	*x 0.57*
newton per square meter	pascal	x 1^	
newton-meter	dyne-centimeters	x 1^E+07^	
	kilogram (force)-meters	x 0.101 97	*x 0.1*
	ounce (force)-inches	x 141.61	*x 140*
	pound (force)-feet	x 0.737 56	*x 0.74*
	pound (force)-inches	x 8.850 7	*x 9*
nibble	bits	x 4^	
	bytes	x 2^	
night, natural	hours, inequal	x 12	
nip (for wine, Brit.)	bottles (Brit.)	/4^ or x 0.25^	
	liters	x 3/16^ or x 0.187 5^	*x 0.19*
nit	stilb	x 1^	
nonary	novenary	x 1^	
nonillion (Brit.)	septendecillion (US)	x 1^	
nonillion (US)	1^E+30	x 1^	
nonuple (quantity)	items	x 9^	
nonuple (size)	nine times as large	x 1	
normal	perpendicular	x 1^	
normal solution	See "concentration, normal"		
nought	zero	x 1^	
novemdecillion (Brit.)	1^E+114	x 1^	
novemdecillion (US)	1^E+60	x 1^	
novenary	based on the number 9	x 1^	
nuclear magneton	Bohr magneton	/1837 or x 5.443 7 E-04	
number denominations, relative values of (above 100)	See APPENDIX, "numbers"		
number, decimal	percent	x 100^	
nutation (Earth-axis oscillation)	years	x 18.6	*x 18*
nybble	nibble	x 1^	
-O-			
obtuse angle	angle greater than 90 degrees	x 1	
octad	group of eight	x 1^	
octal	based on the number 8	x 1^	
octane number (gasoline anti-knock rating)	volume percent of isooctane in a standard reference fuel	x 1^	
octane rating	octane number	x 1^	
octant	circle	/8^ or x 0.125^	
octave (for poetry)	lines in stanza	x 8^	
octave (quantity)	group of eight	x 1^	
octave (sound)	interval between two frequencies having ratio of 2^	x 1^	
	number of half-tones	x 12^	
	number of whole tones	x 6^	
octet	group of eight	x 1^	
octillion (Brit.)	quindecillion (US)	x 1^	
octillion (US)	1^E+27	x 1^	
octodecillion (Brit.)	1^E+108	x 1^	
octodecillion (US)	1^E+57	x 1^	

TABLE 2
Measurement Conversions, Alphabetical

All measurement units are US, unless otherwise noted. All number denominations "billion" and higher are US, unless otherwise noted.

CONVERT	TO EQUIVALENT	BY PRECISELY	OR WITHIN ± 5.0 %
octodenary	based on the number 18	x 1^	
octonary (number)	octal	x 1^	
octonary (quantity)	group of eight	x 1^	
octuple (quantity)	items	x 8^	
octuple (size)	eight times as large	x 1	
oersted	abampere-turns per centimeter	/4 pi or x 0.079 577	*x 0.08*
	ampere-turns per centimeter	x 10/4 pi or x 0.795 77	*x 0.8*
	ampere-turns per inch	x 2.021 3	*x 2*
	ampere-turns per meter	x 79.577	*x 80*
	gilbert per centimeter	x 1^	
ohm	1/mho	x 1^	
	abohms	x 1^E+09	
	EMU of resistance	x 1^E+09	
	ESU of resistance	x 1.112 6 E-12	*x 1.1 E-12*
	statohms	x 1.112 6 E-12	*x 1.1 E-12*
ohm per mil-foot	ohm (mil, foot)	x 1^	
ohm (mil, foot)	microhm-centimeters	x 0.166 24	*x 0.17*
	microhm-inches	x 0.065 450	*x 0.065*
	ohm-meters	x 1.662 4 E-09	*x 1.7 E-09*
ohm, absolute	ohms, international	x 0.999 51	*x 1*
ohm-centimeter	ohm-meters	x 0.01^	
ohm-circular mil per foot	ohm-meters	x 1.662 4 E-09	*x 1.7 E-09*
	ohm-square millimeters per meter	x 0.001 662 4	*x 0.001 7*
ohm-meter	microhm-centimeters	x 1^E+08	
	microhm-inches	x 500 E+07/127^ or x 3.937 0 E+07	*x 4 E+07*
	ohms (mil, foot)	x 6.015 3 E+08	*x 6 E+08*
	ohm-centimeters	x 100^	
	ohm-circular mils per foot	x 6.015 3 E+08	*x 6 E+08*
ohm-square millimeter per meter	ohm-circular mils per foot	x 601.53	*x 600*
order(s) of magnitude	ratio of values of two similar items [14]	x 10^ to power of the no. of orders of magnitude	
ought	zero	x 1^	
ounce	ounce, avoirdupois	x 1^	
ounce of carbohydrate	kilocalories [5]	x 110 (approx.)	
ounce of fat	kilocalories [5]	x 260 (approx.)	
ounce of liquor, 80 proof	kilocalories [5]	x 65	
ounce of protein	kilocalories [5]	x 110 (approx.)	
ounce per cubic inch	kilograms per cubic meter	x 1 730.0	*x 1 700*
ounce per gallon	kilograms per cubic meter	x 7.489 2	*x 7.5*
ounce per Imperial gallon (Brit.)	kilograms per cubic meter	x 6.236 0	*x 6*
ounce (force)	newtons	x 0.278 01	*x 0.28*
ounce (force) per square foot	kilograms (force) per square meter	x 0.305 15	*x 0.3*
ounce (force) per square inch	grams (force) per square centimeter	x 4.394 2	*x 4.4*
ounce (force) per square yard	kilograms (force) per square meter	x 0.033 906	*x 0.034*
ounce (force)-inch	newton-meters	x 0.007 061 6	*x 0.007*

[5] The energy value of food and drink is customarily stated in "calories", but the technically correct measuring unit is "kilocalories", which is sometimes called "large calories". (1 kilocalorie = 1 000 calories) A person on a reducing diet may take 1 100 "calories" per day, but, in truth, he is taking 1 100 kilocalories, or 1.1 million (1 100 000) calories, or 1.1 megacalories.

[14] Orders of magnitude vary by a factor of 10. Examples: one order of magnitude = ratio of 10:1; two orders, 100:1; three orders, 1000:1; etc.

TABLE 2
Measurement Conversions, Alphabetical

All measurement units are US, unless otherwise noted. All number denominations "billion" and higher are US, unless otherwise noted.

CONVERT	TO EQUIVALENT	BY PRECISELY	OR WITHIN ± 5.0 %
ounce, apothecary	carats	x 155.52	*x 160*
	drams, apothecary	x 8^	
	drams, avoirdupois	x 3 072/175^ or x 17.554	*x 17*
	grains	x 480^	
	grams	x 31.103	*x 30*
	kilograms	x 0.031 103	*x 0.03*
	milligrams	x 31 103	*x 30 000*
	ounces, avoirdupois	x 1.097 1	*x 1.1*
	pennyweights	x 20^	
	pounds, apothecary	/12^ or x 0.083 333	*x 0.08*
	pounds, avoirdupois	x 12/175^ or x 0.068 571	*x 0.07*
	pounds, troy	/12^ or x 0.083 333	*x 0.08*
	scruples, apothecary	x 24^	
ounce, apothecary (Brit.)	ounce, apothecary (US)	x 1^	
ounce, apothecary (Can.)	ounce, apothecary (US)	x 1^	
ounce, avoirdupois	carats	x 141.75	*x 140*
	drams, apothecary	x 175/24^ or x 7.291 7	*x 7*
	drams, avoirdupois	x 16^	
	grains	x 437.5^	*x 440*
	grams	x 28.350	*x 28*
	hundredweights, short	/1 600^ or x 6.25^E-04	*x 6 E-04*
	kilograms	x 0.028 350	*x 0.028*
	milligrams	x 28 350	*x 28 000*
	ounces, apothecary	x 175/192^ or x 0.911 46	*x 0.9*
	ounces, troy	x 175/192^ or x 0.911 46	*x 0.9*
	pennyweights	x 875/48^ or x 18.229	*x 18*
	pounds, apothecary	x 175/2 304^ or x 0.075 955	*x 0.076*
	pounds, avoirdupois	/16^ or x 0.062 5^	*x 0.06*
	pounds, troy	x 175/2 304^ or x 0.075 955	*x 0.08*
	scruples, apothecary	x 175/8^ or x 21.875^	*x 22*
	tons, long	x 2.790 2 E-05	*x 2.8 E-05*
	tons, metric	x 2.83 50 E-05	*x 2.8 E-05*
	tons, short	/32 000^ or x 3.125^E-05	*x 3 E-05*
ounce, avoirdupois (Brit.)	ounce, avoirdupois (US)	x 1^	
ounce, avoirdupois (Can.)	ounce, avoirdupois (US)	x 1^	
ounce, avoirdupois (US)	ounce, avoirdupois (Brit.)	x 1^	
ounce, fluid	cubic feet	x 0.001 044 4	*x 0.001*
	cubic inches	x 1.804 7	*x 1.8*
	cubic meters	x 2.957 4 E-05	*x 3 E-05*
	cups, measuring	/8^ or x 0.125^	*x 0.13*
	drams, fluid	x 8^	
	gallons	/128^ or x 0.007 812 5^	*x 0.008*
	gills	/4^ or x 0.25^	
	liters	x 0.029 574	*x 0.03*
	milliliters	x 29.574	*x 30*
	minims	x 480^	*x 500*
	ounces, Imperial fluid (Brit.)	x 1.040 8	*x 1*
	pints, fluid	x 16^ or x 0.062 5^	*x 0.06*
	quarts, fluid	/32^ or x 0.031 25^	*x 0.03*

TABLE 2
Measurement Conversions, Alphabetical

All measurement units are US, unless otherwise noted. All number denominations "billion" and higher are US, unless otherwise noted.

CONVERT	TO EQUIVALENT	BY PRECISELY	OR WITHIN ± 5.0 %
ounce, fluid apothecary	ounce, fluid	x 1^	
ounce, fluid (Can.)	ounce, Imperial fluid (Brit.)	x 1^	
ounce, Imperial fluid apothecary (Brit.)	ounce, Imperial fluid (Brit.)	x 1^	
ounce, Imperial fluid (Brit.)	cubic meters	x 2.841 3 E-05	*x 2.8 E-05*
	drachms, Imperial fluid (Brit.)	x 8^	
	milliliters	x 28.413	*x 28*
	ounces, fluid (US)	x 0.960 76	*x 1*
ounce, troy (Can.)	ounce, troy (US)	x 1^	
ounce, troy (for precious metals)	carats	x 155.52	*x 160*
	drams, apothecary	x 8^	
	drams, avoirdupois	x 3 072/175^ or x 17.554	*x 18*
	grains	x 480^	
	grams	x 31.103	*x 30*
	kilograms	x 0.031 103	*x 0.03*
	milligrams	x 31 103	*x 30 000*
	ounces, avoirdupois	x 192/175^ or x 1.097 1	*x 1.1*
	pennyweights	x 20^	
	pounds, apothecary	/12^ or x 0.083 333	*x 0.08*
	pounds, avoirdupois	x 12/175^ or x 0.068 571	
	pounds, troy	/12^ or x 0.083 333	*x 0.08*
	scruples, apothecary	x 24^	
ounce, troy (for precious metals, Brit.)	ounce, troy (US)	x 1^	
-P-			
pace, double-time	inches	x 36	
pace, geometrical	feet	x 5	
pace, military	inches	x 30	
pace, quick-time	inches	x 30	
pair	related items	x 2^	
palm	inches	x 3	
parsec	astronomical units	x 2.062 6 E+05	*x 2 E+05*
	kilometers	x 3.085 7 E+13	*x 3 E+13*
	light-years	x 3.261 6	*x 3.3*
	meters	x 3.085 7 E+16	*x 3 E+16*
	miles	x 1.917 3 E+13	*x 2 E+13*
part per million (by weight)	grains per gallon (of water @ 20 deg. C)	x 0.058 313	*x 0.06*
	grains per gallon (of water @ 4 deg. C)	x 0.058 417	*x 0.06*
	grains per Imperial gallon (of water @ 20 deg. C, Brit.)	x 0.070 032	*x 0.07*
	grains per Imperial gallon (of water @ 4 deg. C, Brit.)	x 0.070 155	*x 0.07*
	grams per liter (of water @ 4 deg. C)	x 0.001^	
	pounds per gallon (of water @ 20 deg. C)	x 8.330 5 E-06	*x 8 E-06*
	pounds per gallon (of water @ 4 deg. C)	x 8.345 2 E-06	*x 8 E-06*
	pounds per Imperial gallon (of water @ 20 deg. C, Brit.)	x 1.000 5 E-05	*x 1 E-05*
	pounds per Imperial gallon (of water @ 4 deg. C, Brit.)	x 1.002 2 E-05	*x 1 E-05*
	pounds per million gallons (of water @ 20 deg. C)	x 8.330 5	*x 8*

TABLE 2
Measurement Conversions, Alphabetical

All measurement units are US, unless otherwise noted. All number denominations "billion" and higher are US, unless otherwise noted.

CONVERT	TO EQUIVALENT	BY PRECISELY	OR WITHIN ± 5.0 %
	pounds per million gallons (of water @ 4 deg. C)	x 8.345 2	*x 8*
	pounds per million Imperial gallons (of water @ 20 deg. C, Brit.)	x 10.005	*x 10*
	pounds per million Imperial gallons (of water @ 4 deg. C, Brit.)	x 10.022	*x 10*
particle size of clouds	microns	x 0.1 to x 10	
particle size of dusts	microns	x above 10	
particle size of molecules	microns	x below 0.001	
particle size of smokes	microns	x 0.001 to x 0.1	
pascal	atmospheres, standard	x 9.869 2 E-06	*x 10 E-06*
	atmospheres, technical	x 1.019 7 E-05	*x 1 E-05*
	bars	x 1^E-05	
	microbars	x 10^	
	newton per square meter	x 1^	
	pounds (force) per square inch	x 1.450 4 E-04	*x 1.5 E-04*
	torrs	x 0.007 500 6	*x 0.007 5*
pascal absolute	pascals gage	- ambient pressure [11]	
pascal gage	pascals absolute	+ ambient pressure [11]	
pascal-second	centipoises	x 1 000^	
pearl grain (for pearls)	carat grain	x 1^	
peck	bushels	/4^ or x 0.25^	
	cubic feet	x 0.311 11	*x 0.31*
	cubic inches	x 537.61	*x 540*
	cubic meters	x 0.008 809 8	*x 0.009*
	liters	x 8.809 8	*x 9*
	pints, dry	x 16^	
	quarts, dry	x 8^	
peck (Can.)	peck, Imperial (Brit.)	x 1^	
peck, Imperial (Brit.)	cubic inches	x 554.84	*x 550*
	gallons, Imperial (Brit.)	x 2^	
	liters	x 9.092 2	*x 9*
pennyweight	drams, apothecary	x 0.4^	
	drams, avoirdupois	x 0.877 71	*x 0.9*
	grains	x 24^	
	grams	x 1.555 2	*x 1.6*
	kilograms	x 0.001 555 2	*x 0.001 6*
	milligrams	x 1 555.2	*x 1 600*
	ounces, apothecary	/20^ or x 0.05^	
	ounces, avoirdupois	x 48/875^ or x 0.054 857	*x 0.055*
	ounces, troy	/20^ or x 0.05^	
	pounds, apothecary	/240^ or x 0.004 166 7	*x 0.004*
	pounds, avoirdupois	x 3/875^ or x 0.003 428 6	*x 0.003 4*
	pounds, troy	/240^ or x 0.004 166 7	*x 0.004*
	scruples, apothecary	x 1.2^	
pennyweight (Can.)	pennyweight (US)	x 1^	
pennyweight, troy (Brit.)	pennyweight, troy (US)	x 1^	
percent	decimal number	/100^ or x 0.01^	
percent of gold in alloy (by weight)	karats [2]	x 6/25^ or x 0.24^	

2 100-percent pure gold has 24 karats of gold.

11 Ambient pressure is the environmental pressure surrounding a device and is not necessarily atmospheric pressure, whether standard or local.

TABLE 2
Measurement Conversions, Alphabetical

All measurement units are US, unless otherwise noted. All number denominations "billion" and higher are US, unless otherwise noted.

CONVERT	TO EQUIVALENT	BY PRECISELY	OR WITHIN ± 5.0 %
	parts per million (by weight)	x 10 000	
percent solution (by mass)	grams of solute per 100 grams of solution	x 1^	
percent solution (by volume) [3]	liters of solute per 100 liters of solution	x 1^	
percentage point	the difference between two percentages, e.g., 40% minus 32 % = 8 percentage points	x 1^	
perch (Can.)	perch (US)	x 1^	
perch (for land, Brit.)	square rod	x 1^	
perch (for masonry)	cubic feet	x 16.5 or x 24.75 or x 25	
perch (French land area, Queb.)	square feet (French land measure, Queb.)	x 324^	*x 320*
perch (French land length, Queb.)	feet (French land measure, Queb.)	x 18^	
perch (length)	rod (length)	x 1^	
period	1/frequency	x 1^	
period, draconic	nodical month	x 1^	
period, half	See "half-life"		
perm (@ 0 deg. C, for water vapor)	kilograms per pascal-second-square meter	x 5.721 4 E-11	*x 6 E-11*
perm (@ 23 deg. C, for water vapor)	kilograms per pascal-second-square meter	x 5.745 3 E-11	*x 6 E-11*
permeability (magnetic)	1/reluctivity	x 1^	
	henry per meter	x 1^	
permeance	1/reluctance	x 1^	
permittivity	farad per meter	x 1^	
perm-inch (@ 0 deg. C, for water vapor)	kilograms per pascal-second-meter	x 1.453 2 E-12	*x 1.5 E-12*
perm-inch (@ 23 deg. C, for water vapor)	kilograms per pascal-second-meter	x 1.459 3 E-12	*x 1.5 E-12*
perpendicular	at right angles	x 1^	
Petrograd standard (for sawed timber, Can.)	cubic feet	x 165	
pH (= power of hydrogen ion)	negative log of hydrogen-ion activity in gram-ions per liter	x 1^	
phon	sound-pressure decibels of 1 000-hertz tone matching loudness of another sound	x 1^	
phot	lumen per square centimeter	x 1^	
	lumens per square meter	x 10 000^	
	lux	x 10 000^	
photon	quantum of electromagnetic energy	x 1^	
pi	3.141 593	x 1	*355/113; 22/7; or 3*
pica (printer's)	inches	x 0.166 02	*/6 or x 0.17*
	millimeters	x 4.216 9	*x 4.2*
	points (printer's)	x 12^	
pica (typewriter type)	characters per inch	x 10^	
pint (Can.)	pint, Imperial (Brit.)	x 1^	
pint, dry	bushels	/64^ or x 0.015 625^	*x 0.016*
	cubic feet	x 0.019 445	*x 0.02*
	cubic inches	x 33.600	*x 34*
	cubic meters	x 5.506 1 E-04	*x 5.5 E-04*
	liters	x 0.550 61	*x 0.55*

[3] The volume of a solution may differ from the sum of the separate volumes of the solute and solvent.

TABLE 2
Measurement Conversions, Alphabetical

All measurement units are US, unless otherwise noted. All number denominations "billion" and higher are US, unless otherwise noted.

CONVERT	TO EQUIVALENT	BY PRECISELY	OR WITHIN ± 5.0 %
	milliliters	x 550.61	*x 550*
	pecks	/16^ or x 0.062 5^	*x 0.06*
	quarts, dry	/2^ or x 0.5^	
pint, fluid	cubic feet	x 0.016 710	*x 0.017*
	cubic inches	x 28.875^	*x 29*
	cups	x 2^	
	drams, fluid	x 128^	*x 130*
	gallons	/8^ or x 0.125^	*x 0.13*
	gills	x 4^	
	liters	x 0.473 18	*x 0.47*
	milliliters	x 473.18	*x 470*
	minims	x 7 680^	*x 7 700*
	ounces, fluid	x 16^	
	quarts, fluid	/2^ or x 0.5^	
pint, Imperial apothecary (Brit.)	pint, Imperial (Brit.)	x 1^	
pint, Imperial (Brit.)	gills, Imperial (Brit.)	x 4^	
	ounces, Imperial fluid (Brit.)	x 20^	
pIon (= power of any specified ion)	negative log of specified-ion activity in gram-ions per liter	x 1^	
pipe	gallons	x 126 (usually)	
	hogsheads	x 2	
	tuns	/2 or x 0.5	
pitch increase of half tone	See "frequency ratio . . . "		
pitch increase of whole tone	See "frequency ratio . . . "		
point (compass)	circle	/32^ or x 0.031 25^	
	degrees (angle)	x 11.25^	*x 11*
point (jeweler's)	milligrams	x 2^	
	carats	/100^ or x 0.01^	
point (printer's)	inches	x 0.013 835	*x 0.014*
	millimeters	x 350/996^ or x 0.351 41	*x 0.35*
point (thickness of paper)	inches	/1 000^	
point, basis (for loan investment yields)	See "basis point"		
point, percentage	See "percentage point"		
point, quarter (compass)	degrees (angle)	x 45/16^ or x 2.812 5^	*x 2.8*
point, radix (for numbers)	See "radix point"		
poise	centipoises	x 100^	
	dyne-second per square centimeter	x 1^	
	pascal-seconds	/10^ or x 0.1^	
	pound (force)-seconds per square foot	x 0.002 088 5	*x 0.002*
	pounds (mass) per foot-second	x 0.067 197	*x 0.07*
poise (viscosity)	1/rhe (fluidity)	x 1^	
pole (Can.)	pole (US)	x 1^	
pole (length)	rod (length)	x 1^	
poncelet	foot-pounds (force) per second	x 723.30	*x 720*
	horsepower	x 1.315 1	*x 1.3*
	horsepower, metric	x 4/3^ or x 1.333 3	*x 1.3*
	kilocalories (IT) per second	x 0.234 23	*x 0.23*
	kilowatts	x 0.980 66	*x 1*
	meter-kilograms (force) per second	x 100^	

TABLE 2
Measurement Conversions, Alphabetical

All measurement units are US, unless otherwise noted. All number denominations "billion" and higher are US, unless otherwise noted.

CONVERT	TO EQUIVALENT	BY PRECISELY	OR WITHIN ± 5.0 %
	watts	x 980.66	*x 1 000*
pottle (Brit.)	gallons, Imperial (Brit.)	/2 or x 0.5	
pound	pound, avoirdupois	x 1^	
pound of water per minute (@ 4 deg. C)	cubic feet per second	x 2.669 8 E-04	*x 2.7 E-04*
pound per cubic foot	grams per milliliter	x 0.016 018	*x 0.016*
	kilograms per cubic meter	x 16.018	*x 16*
	pounds per cubic inch	x 5.787 0 E-04	*x 6 E-04*
	pounds per gallon	x 0.133 68	*x 0.13*
	tons, short per cubic yard	x 0.013 5^	*x 0.013*
pound per cubic inch	grams per milliliter	x 27.680	*x 28*
	pounds per cubic foot	x 1 728^	*x 1 700*
	pounds per gallon	x 231^	*x 230*
	tons, short per cubic yard	x 23.328^	*x 23*
pound per cubic yard	kilograms per cubic meter	x 0.593 28	*x 0.6*
pound per gallon	grains per cubic inch	x 1 000/33^ or x 30.303	*x 30*
	kilograms per cubic meter	x 27 680	*x 28 000*
	grams per liter	x 0.119 83	*x 0.12*
	kilograms per cubic meter	x 119.83	*x 120*
	ounces per cubic inch	x 0.069 264	*x 0.07*
	pounds per cubic foot	x 7.480 5	*x 7.5*
	tons, short per cubic yard	x 0.100 99	*x 0.1*
	grains per gallon	x 7 000^	
	pounds per million gallons	x 1^E+06	
pound per gallon (of water @ 20 deg. C)	parts per million (by weight)	x 1.200 4 E+05	*x 1.2 E+05*
pound per gallon (of water @ 4 deg. C)	parts per million (by weight)	x 1.198 3 E+05	*x 1.2 E+05*
pound per hour	kilograms per second	x 1.259 9 E-04	*x 1.3 E-04*
pound per Imperial gallon (Brit.)	grains per Imperial gallon	x 7 000^	
	kilograms per cubic meter	x 99.776	*x 100*
pound per Imperial gallon (of water @ 20 deg. C)	parts per million (by weight)	x 99 955	*x 1 E+05*
pound per Imperial gallon (of water @ 4 deg. C)	parts per million (by weight)	x 99 780	*x 1 E+05*
pound per inch	grams per meter	x 17 858	*x 18 000*
pound per million gallons	grains per gallon	x 0.007^	
	pounds per gallon	x 1^E-06	
pound per million gallons (of water @ 20 deg. C)	parts per million (by weight)	x 0.120 04	*x 0.12*
pound per million gallons (of water @ 4 deg. C)	parts per million (by weight)	x 0.119 83	*x 0.12*
pound per million Imperial gallons (Brit.)	grains per Imperial gallon (Brit.)	x 0.007^	
pound per million Imperial gallons (of water @ 20 deg. C, Brit.)	parts per million (by weight)	x 0.099 955	*x 0.1*
pound per million Imperial gallons (of water @ 4 deg. C, Brit.)	parts per million (by weight)	x 0.099 779	*x 0.1*

TABLE 2
Measurement Conversions, Alphabetical

All measurement units are US, unless otherwise noted. All number denominations "billion" and higher are US, unless otherwise noted.

CONVERT	TO EQUIVALENT	BY PRECISELY	OR WITHIN ± 5.0 %
pound per minute	kilograms per second	x 0.007 559 9	x 0.007 6
pound per second	kilograms per second	x 0.453 59	x 0.45
pound (force)	dynes	x 4.448 2 E+05	x 4.4 E+05
	grams (force)	x 453.59	x 450
	kips	/1 000^ or x 0.001^	
	newtons	x 4.448 2	x 4.4
	poundals	x 32.174	x 32
pound (force) per foot	kilograms (force) per meter	x 1.488 2	x 1.5
	newtons per meter	x 14.594	x 15
pound (force) per inch	newtons per meter	x 175.13	x 180
pound (force) per square foot	kilograms (force) per square meter	x 4.882 4	x 5
	pascals	x 47.880	x 50
	pounds (force) per square inch	/144^ or x 0.006 944 4	x 0.007
pound (force) per square inch	atmospheres, standard	x 0.068 046	x 0.07
	bars	x 0.068 948	x 0.07
	dynes per square centimeter	x 68 948	x 70 000
	feet of water (@ 60 deg. F)	x 2.309 0	x 2.3
	grams (force) per square centimeter	x 70.307	x 70
	inches of mercury (@ 60 deg. F)	x 2.041 8	x 2
	inches of water (@ 60 deg. F)	x 27.708	x 28
	kilograms (force) per square centimeter	x 0.070 307	x 0.07
	kilograms (force) per square meter	x 703.07	x 700
	kilopascals	x 6.894 8	x 7
	kips per square inch	/1 000^ or x 0.001^	
	millibars	x 68.948	x 70
	pascals	x 6 894.8	x 7 000
	poundals per square foot	x 4 633.1	x 4 600
	pounds (force) per square foot	x 144^	
	torrs	x 51.715	x 50
pound (force) per square inch absolute	pounds (force) per square inch gage	- ambient pressure [11]	
pound (force) per square inch gage	pounds (force) per square inch absolute	+ ambient pressure [11]	
pound (force)-foot	newton-meters	x 1.355 8	x 1.4
pound (force)-foot per inch	newton-meters per meter	x 53.379	x 53
pound (force)-inch	newton-meters	x 0.112 98	x 0.11
pound (force)-inch per inch	newton-meters per meter	x 4.448 2	x 4/9
pound (force)-second per square foot	centipoises	x 47 880	x 48 000
	pascal-seconds	x 47.880	x 48
	poises	x 478.80	x 480
	pounds (mass) per foot-second	x 32.174	x 32
	slug per foot-second	x 1^	
pound (force)-second per square inch	pascal-seconds	x 6 894.8	x 7 000
pound (mass) per foot-hour	pascal-seconds	x 4.133 8 E-04	x 4 E-04
pound (mass) per foot-second	centipoises	x 1 488.2	x 1 500
	pascal-seconds	x 1.488 2	x 1.5

[11] Ambient pressure is the environmental pressure surrounding a device and is not necessarily atmospheric pressure, whether standard or local.

TABLE 2
Measurement Conversions, Alphabetical

All measurement units are US, unless otherwise noted. All number denominations "billion" and higher are US, unless otherwise noted.

CONVERT	TO EQUIVALENT	BY PRECISELY	OR WITHIN ± 5.0 %
	poises	x 14.882	*x 15*
	pound (force)-seconds per square foot	x 0.031 081	*x 0.03*
poundal	grams (force)	x 14.098	*x 14*
	newtons	x 0.138 26	*x 0.14*
	pound (mass)-foot per second per second	x 1^	
	pounds (force)	x 0.031 081	*x 0.03*
poundal per square foot	pascals	x 1.488 2	*x 1.5*
	pounds (force) per square inch	x 2.158 4 E-04	*x 2.2 E-04*
poundal-second per square foot	pascal-seconds	x 1.488 2	*x 1.5*
pound, apothecary	drams, apothecary	x 96^	
	drams, avoirdupois	x 36 864/175^ or x 210.65	*x 210*
	grains	x 5 760^	*x 6 000*
	grams	x 373.24	*x 370*
	kilograms	x 0.373 24	*x 0.37*
	milligrams	x 3.732 4 E+05	*x 3.7 E+05*
	ounces, apothecary	x 12^	
	ounces, avoirdupois	x 2 304/175^ or x 13.166	*x 13*
	ounces, troy	x 12^	
	pennyweights	x 240^	
	pounds, avoirdupois	x 144/175^ or x 0.822 86	*x 0.8*
	scruples, apothecary	x 288^	*x 300*
pound, apothecary (Brit.)	pound, apothecary (US)	x 1^	
pound, apothecary (Can.)	pound, apothecary (US)	x 1^	
pound, avoirdupois	drams, apothecary	x 350/3^ or x 116.67	*x 120*
	drams, avoirdupois	x 256^	*x 260*
	grains	x 7 000^	
	grams	x 453.59	*x 450*
	hundredweights, short	/100^ or x 0.01^	
	kilograms	x 0.453 59	*x 11/24 or x 4/9*
	milligrams	x 4.535 9 E+05	*x 4.5 E+05*
	ounces, apothecary	x 175/12^ or x 14.583	*x 15*
	ounces, avoirdupois	x 16^	
	ounces, troy	x 175/12^ or x 14.583	*x 15*
	pennyweights	x 875/3^ or x 291.67	*x 300*
	pound (customary)	x 1^	
	pounds, apothecary	x 175/144^ or x 1.215 3	*x 1.2*
	pounds, troy	x 175/144^ or x 1.215 3	*x 1.2*
	scruples, apothecary	x 350^	
	tons, long	/2 240^ or x 4.464 3 E-04	*x 4.4 E-04*
	tons, metric	x 4.535 9 E-04	*x 4.5 E-04*
	tons, short	/2 000^ or x 5^E-04	
pound, avoirdupois (Brit.)	pound, avoirdupois (US)	x 1^	
pound, avoirdupois (Can.)	pound, avoirdupois (US)	x 1^	
pound, troy	drams, apothecary	x 96^	
	drams, avoirdupois	x 36 864/175^ or x 210.65	*x 210*
	grains	x 5 760^	*x 6 000*
	grams	x 373.24	*x 370*
	kilograms	x 0.373 24	*x 0.37*

TABLE 2
Measurement Conversions, Alphabetical

All measurement units are US, unless otherwise noted. All number denominations "billion" and higher are US, unless otherwise noted.

CONVERT	TO EQUIVALENT	BY PRECISELY	OR WITHIN ± 5.0 %
	ounces, apothecary	x 12^	
	ounces, avoirdupois	x 2 304/175^ or x 13.166	*x 13*
	ounces, troy	x 12^	
	pennyweights	x 240^	
	pounds, avoirdupois	x 144/175^ or x 0.822 86	*x 0.8*
	scruples, apothecary	x 288^	*x 300*
pound, troy (Brit.)	pound, troy (US)	x 1^	
pound, troy (Can.)	pound, troy (US)	x 1^	
pound-atom	mass of element, in pounds, equal in number to atomic weight	x 1^	
pound-molecule	mass of molecule, in pounds, equal in number to molecular weight	x 1^	
pound-square foot	gram-square centimeters	x 4.214 0 E+05	*x 4.2 E+05*
	kilogram-square meters	x 0.042 140	*x 0.042*
	pound-square inches	x 144^	
	slug-square feet	x 0.031 081	*x 0.03*
pound-square inch	gram-square centimeters	x 2 926.4	*x 3 000*
	kilogram-square meters	x 2.926 4 E-04	*x 3 E-04*
	pound-square feet	/144^ or x 0.006 944 4	*x 0.007*
	slug-square feet	x 2.158 4 E-04	*x 2.2 E-04*
power density	watt per square meter	x 1^	
power, optical	magnification	x 1^	
ppb	parts per billion	x 1^	
ppm (concentration)	parts per million	x 1^	
ppm (for computers)	pages (8.5 x 11 inches, double-spaced) per minute	x 1^	
proof spirit (Brit.)	57.10 percent ethanol (by volume) in distilled water	x 1^	
proof spirit (U.S. standard concentration)	50 percent ethanol (by volume) in water (@ 60 deg. F)	x 1^	
proof (for liquors)	ethanol percent (by volume) in water	x 50	
proton rest mass	atomic mass units	x 1.007 3	*x 1*
psi	pascals	x 6 894.8	*x 7 000*
	pounds (force) per square inch	x 1^	
psia	pound (force) per square inch, absolute	x 1^	
psid	pound (force) per square inch, differential	x 1^	
	psi	x 1^	
psig	pound (force) per square inch, gage	x 1^	
puncheon	gallons	x 84 (usually)	
puncheon (Brit.)	barrels, Imperial (Brit.)	x 2	
	gallons, Imperial (Brit.)	x 72 (usually)	
-Q-			
quad (area)	quadrangle	x 1^	
quad (energy)	Btu (IT)	x 1^E+15	
	joules	x 1.055 E+18	
quad (quantity)	group of four	x 1^	
quadragenary	based on the number 40	x 1^	
quadragesimal	group of 40	x 1^	
quadrant (angle)	degrees (angle)	x 90^	
	minutes (angle)	x 5 400^	
	quarter circle	x 1^	
	radians	x 1.570 8	*x 1.6*
	revolutions	/4^ or x 0.25^	

TABLE 2
Measurement Conversions, Alphabetical

All measurement units are US, unless otherwise noted. All number denominations "billion" and higher are US, unless otherwise noted.

CONVERT	TO EQUIVALENT	BY PRECISELY	OR WITHIN ± 5.0 %
	seconds (angle)	x 3.24^E+05	*x 3.2 E+05*
quadrennial	years	x 4^	
quadrennium	years	x 4^	
quadrilateral	sides	x 4^	
quadrillion (Brit.)	septillion (US)	x 1^	
quadrillion (US)	1^E+15	x 1^	
quadripartite	parts	x 4^	
quadruple (quantity)	items	x 4^	
quadruple (size)	four times as large	x 1^	
quantum of charge	coulombs	x 1.602 1 E-19	*x 1.6 E-19*
quart (Can.)	quart, Imperial (Brit.)	x 1^	
quarter circle	degrees (angle)	x 90^	
	quadrant	x 1^	
quarter section	acres	x 160^	
	forties	x 4^	
quarter (Brit.)	pounds, avoirdupois	x 28^	
	stones (Brit.)	x 2^	
quarter (length)	furlongs	x 2^	
	miles	/4^ or x 0.25^	
quarter (quantity)	item	/4^	
quarter (size)	the whole	/4^ or x 0.25^	
quarter (time)	months	x 3	
	school years	/4^	
quartern (number)	a fourth	x 1^	
quartern (quantity)	parts	/4 or x 0.25	
quartern (for a loaf of bread, Brit.)	pounds	x 4	
quarter, Imperial (volume, Brit.)	bushels, Imperial (Brit.)	x 8^	
quartet	group of four	x 1^	
quart, dry	bushels, struck measure	/32^ or x 0.031 25^	*x 0.03*
	cubic feet	x 0.038 889	*x 0.04*
	cubic inches	x 67.201	*x 67*
	cubic meters	x 0.001 101 2	*x 0.001 1*
	liters	x 1.101 2	*x 1.1*
	pecks	/8^ or x 0.125^	*x 0.13*
	pints, dry	x 2^	
quart, dry (US)	quarts, Imperial (Brit.)	x 0.968 94	*x 0.97*
quart, fluid	cubic feet	x 0.033 420	*x 0.033*
	cubic inches	x 57.75^	*x 60*
	drams, fluid	x 256^	*x 260*
	gallons	/4^ or x 0.25^	
	gills	x 8^	
	liters	x 0.946 35	*x 0.95*
	milliliters	x 946.35	*x 950*
	minims	x 15 360^	*x 15 000*
	ounces, fluid	x 32^	
	pints, fluid	x 2^	
	quarts, dry	x 0.859 37	*x 0.9*
quart, fluid (US)	quarts, Imperial (Brit.)	x 0.832 67	*x 0.8*
quart, Imperial (Brit.)	ounces, Imperial fluid (Brit.)	x 40^	
	pints, Imperial (Brit.)	x 2^	
	quarts, dry (US)	x 1.032 1	*x 1*

TABLE 2
Measurement Conversions, Alphabetical

All measurement units are US, unless otherwise noted. All number denominations "billion" and higher are US, unless otherwise noted.

CONVERT	TO EQUIVALENT	BY PRECISELY	OR WITHIN ±5.0 %
	quarts, fluid (US)	x 1.200 9	*x 1.2*
quaterdenary	based on the number 14	x 1^	
quaternary (number)	based on the number 4	x 1^	
quaternary (quantity)	group of four	x 1^	
quattuordecillion (Brit.)	1^E+84	x 1^	
quattuordecillion (US)	1^E+45	x 1^	
quinary (number)	based on the number 5	x 1^	
quinary (quantity)	group of five	x 5^	
quinate	arranged in groups of five	x 1^	
quindecennial	years	x 15^	
quindecillion (Brit.)	1^E+90	x 1^	
quindecillion (US)	1^E+48	x 1^	
quinquagesimal	days	x 50^	
quinquennium	years	x 5^	
quintal (Brit.)	pounds, avoirdupois	x 112^	*x 110*
quintal, long	hundredweight, long	x 1^	
quintal, metric	kilograms	x 100^	
	pounds, avoirdupois	x 220.46	*x 220*
quintal, short	hundredweight, short	x 1^	
quintan	occurring every fifth day	x 1^	
quintet	group of five	x 1^	
quintillion (Brit.)	nonillion (US)	x 1^	
quintillion (US)	1^E+18	x 1^	
quintuple (quantity)	items	x 5^	
quintuple (size)	five times as large	x 1^	
quire	reams	/20^	
	sheets of paper	x 25^ (usual) or x 24^	
quotidian	daily	x 1^	
-R-			
rad (absorbed dose of radiation))	ergs per gram (of absorbed radiation energy)	x 100	
	grays	x 0.01^	
rad (angle) per second	radian per second	x 1^	
radian	57 deg. 17 min. 44.806 sec.	x 1	
	degrees (angle)	x 180/pi or x 57.296	*x 60*
	grades	x 63.662	*x 64*
	mils (angle)	x 1 018.6	*x 1 000*
	minutes (angle)	x 10 800/pi or x 3 437.7	*x 3 500*
	quadrant (angle)	x 0.636 62	*x 0.64*
	revolutions	/2 pi or x 0.159 15	*x 0.16*
	seconds (angle)	x 6.48 E+05/pi or x 2.062 6 E+05	*x 2 E+05*
radian per second	degrees (angle) per second	x 57.296	*x 57*
	revolutions per minute	x 9.549 3	*x 10*
	revolutions per second	x 0.159 15	*x 0.16*
radian per second per second	revolutions per minute per minute	x 572.96	*x 600*
	revolutions per minute per second	x 9.549 3	*x 10*
	revolutions per second per second	x 0.159 15	*x 0.16*
radiance	watt per square meter-steradian	x 1^	
radiant intensity	watt per steradian	x 1^	
radix	the base of a number system	x 1^	
	the number of symbol types in a number system	x 1^	

TABLE 2
Measurement Conversions, Alphabetical

All measurement units are US, unless otherwise noted. All number denominations "billion" and higher are US, unless otherwise noted.

CONVERT	TO EQUIVALENT	BY PRECISELY	OR WITHIN ± 5.0 %
radix point	the character separating the integer and fraction parts of a number	x 1^	
railway track gage, narrow	inches	x 36^ (usually), 30^, or 42^	
railway track gage, standard	inches	x 56.5^	*x 57*
rain, inch of	See "inch of rain"		
range, audible sound	hertz	x 20 to x 20 000	
range, human sound sensitivity (power, minimum up to pain)	watts per square centimeter	x 1 E-16 to x 0.01	
range, human sound sensitivity (pressure, minimum up to pain)	dynes per square centimeter	x 2 E-04 to x 2 000	
range, infrasound	hertz	x below 20	
range, ultrasound	hertz	x above 20 000	
ratio of measured sound power to reference power [16]	decibels	x 10 log of ratio	
ratio of measured sound pressure to reference pressure [17]	decibels	x 20 log of ratio	
ratio of two electric current levels having equal resistances	decibels	x 20 log of ratio	
ratio of two electric power levels	decibels	x 10 log of ratio	
	nepers	x 0.5 ln of ratio	
ratio of two voltage levels having equal resistances	decibels	x 20 log of ratio	
ratio, golden (for esthetic design)	0.618 03 or 1.618 03 [10]	x 1	*3/5 or 5/3*
ream	quires	x 20^	
	sheets of drawing or handmade paper	x 472^	*x 470*
ream, long (usual)	sheets of paper	x 500^	
ream, perfect	ream, printer's	x 1^	
ream, printer's	sheets of paper	x 516^	*x 500*
ream, short	sheets of paper	x 480^	*x 500*
rehoboam (for wine, Brit.)	bottles (Brit.)	x 6^	
	liters	x 4.5	
rel	ampere-turn per maxwell	x 1^	
relative density	See "density, relative"		
relative values of number denominations	See APPENDIX, "numbers"		
reluctance	1/permeance	x 1^	
reluctivity	1/permeability (magnetic)		
rem (roentgen equivalent man)	radiation dose causing human biological damage equivalent to that from one roentgen of 200-kV x-rays	x 1	
	roentgen of 200-kV X-rays	x 1^	

[10] The golden ratio equals - 0.5 ± the square root of 1.25.
[16] The reference level for sound power is 1 E-16 watts per square centimeter.
[17] The reference level for sound pressure is 2 E-04 dynes per square centimeter.

TABLE 2
Measurement Conversions, Alphabetical

All measurement units are US, unless otherwise noted. All number denominations "billion" and higher are US, unless otherwise noted.

CONVERT	TO EQUIVALENT	BY PRECISELY	OR WITHIN ±5.0 %
	sieverts	x 0.01^	
rep (roentgen equivalent physical)	energy absorption of 93 ergs per gram in soft tissue	x 1^	
	ergs per gram (of radiation energy absorbed in soft tissue)	x 93	*x 90*
resistance, specific	resistivity	x 1^	
reverberation time	time for a cut-off sound to decrease to one-millionth of its initial intensity	x 1^	
revolution	degrees (angle)	x 360^	
	mils (angle)	x 6 400^	
	quadrants (angle)	x 4^	
	radians	x 2 pi or x 6.283 2	*x 6*
revolution per minute	degrees (angle) per second	x 6^	
	radians per second	x 0.104 72	*x 0.1*
	revolutions per second	/60^ or x 0.016 667	*x 0.017*
revolution per minute per minute	radians per second per second	x 0.001 745 3	*x 0.001 7*
	revolutions per second per second	/3 600^ or x 2.777 8 E-04	*x 2.8 E-04*
revolution per minute per second	radians per second per second	x 0.104 72	*x 0.1*
	revolutions per minute per minute	x 60^	
	revolutions per second per second	/60^ or x 0.016 667	
revolution per second	degrees (angle) per second	x 360^	
	radians per second	x 2 pi or x 6.283 2	*x 6*
	revolutions per minute	x 60^	
revolution per second per second	radians per second per second	x 6.283 2	*x 6*
	revolutions per minute per minute	x 3 600^	
rhe	1/pascal-second (viscosity)	x 1^	
	1/poise (viscosity)	x 1^	
Richter scale (concentration)	ethanol percent (by volume) in water, in deg. Richter	x 1^	
Richter scale (for earthquakes) increase of 0.1 magnitude [23]	freed-energy increase	x 10 to power 0.15 or x 1.4	
	ground-motion-amplitude increase	x 10 to power 0.1 or x 1.258 9	*x 1.3*
	seismic-moment increase	x 10 to power 0.15 or x 1.4	
Richter scale (for earthquakes) increase of one magnitude [23]	freed-energy increase	x 10 to power 1.5 or x 32	
	ground-motion-amplitude increase	x 10	
	seismic-moment increase	x 10 to power 1.5 or x 32	
right angle	degrees (angle)	x 90^	
rms	root mean square	x 1^	
rod (area)	square yards	x 30.25^	*x 30*
rod (length)	chains, surveyor's	/4^ or x 0.25^	
	feet	x 16.5^	*x 17*
	furlongs	/40^ or x 0.025^	
	links, engineer's	x 16.5^	*x 17*
	links, surveyor's	x 25^	
	meters	x 5.029 2	*x 5*
	miles	x 5/1 600^ or x 0.003 125^	*x 0.003 1*
	yards	x 5.5^	

[23] See APPENDIX, "earthquakes".

TABLE 2
Measurement Conversions, Alphabetical

All measurement units are US, unless otherwise noted. All number denominations "billion" and higher are US, unless otherwise noted.

CONVERT	TO EQUIVALENT	BY PRECISELY	OR WITHIN ± 5.0 %
rod (Can.)	rod (US)	x 1^	
rod, cubic (Eng.)	See "cubic rod"		
roentgen	coulombs per kilogram	x 2.58 E-04	*x 2.6 E-04*
Roman number	See APPENDIX, "numbers"		
rood (area, Brit.)	acres	/4^ or x 0.25^	
	square rods	x 40^	
	square yards	x 1 210^	*x 1 200*
rood (Can.)	hectares	x 0.101 17	*x 0.1*
rood (length, Brit.)	yards	x 5.5 to x 8	
root mean square	square root of arithmetic mean of the squares of a group of numbers	x 1	
root-mean-square deviation	standard deviation	x 1^	
rpm	revolution per minute	x 1^	
rutherford	radioactive disintegrations per second	x 1^E+06	
R-value	x 1/thermal conductance	x 1^	
-S-			
Saffir-Simpson hurricane scale	See APPENDIX, "wind speeds"		
salmanazar (for wine, Brit.)	bottles (Brit.)	x 12^	
	liters	x 9.0	
salt cart (Can.)	liters	x 490.98	*x 500*
salt tub (Can.)	salt cart (Can.)	/6^ or x 0.166 67	
score (number)	20^	x 1^	
score (quantity)	items	x 20^	
scruple (Brit.)	drachms (Brit.)	/3^ or x 0.333 33	*x 0.33*
scruple, apothecary	drams	/3^ or x 0.333 33	*x 0.33*
	drams, avoirdupois	x 128/175^ or x 0.731 43	*x 0.7*
	grains	x 20^	
	grams	x 1.296 0	*x 1.3*
	kilograms	x 0.001 296 0	*x 0.001 3*
	milligrams	x 1 296.0	*x 1 300*
	ounces, apothecary	/24^ or x 0.041 667	*x 0.04*
	ounces, avoirdupois	x 8/175^ or x 0.045 714	*x 0.046*
	ounces, troy	/24^ or x 0.041 667	*x 0.04*
	pennyweights	5/6^ or x 0.833 33	*x 0.8*
	pounds, apothecary	/288^ or x 0.003 472 2	*x 0.003 5*
	pounds, avoirdupois	/350^ or x 0.002 857 1	*x 0.002 9*
	pounds, troy	/288^ or x 0.003 472 2	*x 0.003 5*
scruple, apothecary (Brit.)	scruple, apothecary (US)	x 1^	
scruple, Imperial fluid (Brit.)	drachms, Imperial fluid (Brit.)	/3^ or x 0.333 33	*x 0.33*
	minims, Imperial (Brit.)	x 20^	
	ounce, Imperial fluid (Brit.)	/24^ or x 0.041 667	
second (angle)	degrees (angle)	/3 600^ or x 2.777 8 E-04	*x 2.8 E-04*
	minutes (angle)	/60^ or x 0.016 667	*x 0.017*
	radians	x pi/6.48^E+05 or x 4.848 1 E-06	*x 5 E-06*
	revolutions	/1 296^E+06 or x 7.716 0 E-07	*x 8 E-07*
second (customary)	second, mean solar	x 1^	
second, ephemeris	years, tropical	x 3.168 9 E-08	*x 3.2 E-08*
second, mean solar	days, mean solar	/86 400^ or x 1.157 4 E-05	*x 1.2 E-05*
	days, sidereal	x 1.160 6 E-05	*x 1.2 E-05*
	hours, mean solar	/3 600^ or x 2.777 8 E-04	*x 2.8 E-04*

TABLE 2
Measurement Conversions, Alphabetical

All measurement units are US, unless otherwise noted. All number denominations "billion" and higher are US, unless otherwise noted.

CONVERT	TO EQUIVALENT	BY PRECISELY	OR WITHIN ± 5.0 %
	minutes, mean solar	/60^ or x 0.016 667	*x 0.017*
	seconds, sidereal	x 1.002 7	*x 1*
second, sidereal	days, mean solar	x 1.154 2 E-05	*x 1.2 E-05*
	days, sidereal	x 1.157 4 E-05	*x 1.2 E-05*
	hours, sidereal	/3 600^ or x 2.777 8 E-04	*x 2.8 E-04*
	minutes, sidereal	/60^ or x 0.016 667	*x 0.017*
	seconds, mean solar	x 0.997 27	*x 1*
section	square mile	x 1^	
semester	school years	/2^	
semicircle	degrees (angle)	x 180^	
semitone	half-tone	x 1^	
semi-infinite	extending to infinity in one direction or dimension	x 1	
senary	based on the number 6	x 1^	
septenary (number)	based on the number 7	x 1^	
septenary (quantity)	group of seven	x 1^	
septendecillion (Brit.)	1^E+102	x 1^	
septendecillion (US)	1^E+54	x 1^	
septendecimal	based on the number 17	x 1^	
septet	group of seven	x 1^	
septillion (Brit.)	tredecillion (US)	x 1^	
septillion (US)	1^E+24	x 1^	
septuple (quantity)	items	x 7^	
septuple (size)	seven times as large	x 1^	
sesquicentennial	years	x 150^	
sexadecimal	based on the number 16	x 1^	
sexagenary	based on the number 60	x 1^	
sexagesimal	sexagenary	x 1^	
sexdecillion (Brit.)	1^E+96	x 1^	
sexdecillion (US)	1^E+51	x 1^	
sexpartite	parts	x 6^	
sextant	circle	/6^ or x 0.166 67	
sextet	group of six	x 1^	
sextillion (Brit.)	undecillion (US)	x 1^	
sextillion (US)	1^E+21	x 1^	
sextuple (quantity)	items	x 6^	
sextuple (size)	six times as large	x 1^	
shake	seconds	x 1^E-08	
siemens	mho	x 1^	
	abmhos	x 1^E-09	
	statmhos	x 8.987 6 E+11	*x 9 E+11*
sievert	rems (roentgen equivalent man)	x 100^	
sigma (statistical)	standard deviation	x 1^	
sigma (time)	seconds	/1 000^ or x 0.001^	
Sikes scale	ethanol percent (by volume) in water, in deg. Sikes	x 1^	
single	one item	x 1^	
skein	feet	x 360^	
	meters	x 109.73	*x 110*
slug	kilograms	x 14.594	*x 15*
	pound (force) per foot per second per second	x 1^	
slug per cubic foot	kilograms per cubic meter	x 515.38	*x 500*

TABLE 2
Measurement Conversions, Alphabetical

All measurement units are US, unless otherwise noted. All number denominations "billion" and higher are US, unless otherwise noted.

CONVERT	TO EQUIVALENT	BY PRECISELY	OR WITHIN ± 5.0 %
slug, metric	kilogram (force) per meter per second per second	x 1^	
slug-square foot	gram-square centimeters	x 1.355 8 E+07	*x 1.4 E+07*
	kilogram-square meters	x 1.355 8	*x 1.4*
	pound-square feet	x 32.174	*x 32*
	pound-square inches	x 4 633.1	*x 4 600*
solar constant (for radiation falling perpendicularly to the earth's surface)	calories per minute-square centimeter	x 1.94 (mean)	*x 2*
solo	one alone	x 1^	
sone	loudness of 1 000-hertz tone at 40 decibels above listener's audibility threshold	x 1^	
sound intensity	See "sound power level" and "sound pressure level"		
sound power intensity	sound power	x 1^	
sound power level	See "decibel"		
sound pressure amplitude	sound pressure	x 1^	
sound pressure level	micropascals	x 20^	
	See "decibel"		
sound-power increase of 0.1 decibel	sound-power intensity increase	x 10 to power 0.01 or x 1.023 3	*x 1.0*
sound-power increase of one decibel	sound-power intensity increase	x 10 to power 0.1 or x 1.258 9	*x 1.3*
sound-power-level reference	watts per square centimeter	x 1^E-16	
sound-pressure increase of 0.1 decibel	sound-pressure intensity increase	x 10 to power 0.005 or x 1.011 6	*x 1.0*
sound-pressure increase of one decibel	sound-pressure intensity increase	x 10 to power 0.05 or x 1.122 0	*x 1.1*
sound-pressure-level reference	dynes per square centimeter (in air)	x 2^E-04	
	microbars	x 2^E-04	
span	inches	x 9^	
specific charge	ratio of electrical particle charge to its mass	x 1^	
specific energy	joule per kilogram	x 1^	
specific entropy	joule per kilogram-kelvin	x 1^	
specific heat capacity	joule per kilogram-kelvin	x 1^	
specific impulse	ratio of rocket thrust to propellant consumption rate	x 1^	
specific refractivity	ratio of refractivity of a medium to its density	x 1^	
specific surface	ratio of surface area of a powder to its mass	x 1^	
	ratio of total surface of adsorbent substance to its volume	x 1^	
specific thrust	specific impulse	x 1^	
specific volume	1/density	x 1^	
	cubic meter per kilogram	x 1^	
specific weight	kilogram (force) per cubic meter	x 1^	
	pound (force) per cubic foot	x 1^	
	ratio of mass of a substance to its volume	x 1^	
speed of light (in vacuum)	kilometers per second	x 2.997 9 E+05	*x 3 E+05*
	miles per second	x 1.862 8 E+05	
spheradian	steradian	x 1^	

TABLE 2
Measurement Conversions, Alphabetical

All measurement units are US, unless otherwise noted. All number denominations "billion" and higher are US, unless otherwise noted.

CONVERT	TO EQUIVALENT	BY PRECISELY	OR WITHIN ± 5.0 %
sphere	hemispheres	x 2^	
	spherical right angles	x 8^	
	steradians	x 4 pi or x 12.566	*x 13*
spherical candlepower	lumens	x 4 pi or x 12.566	*x 13*
spherical degree	solid right angles	/90^ or x 0.011 111	*x 0.011*
spherical right angle	hemispheres	/4^ or x 0.25^	
	spheres	/8^ or x 0.125^	
	steradians	x pi/2 or x 1.570 8	*x 1.6*
spherical solid angle	steradians	x 4 pi or x 12.566	*x 13*
split (for beverage containers)	ounces, fluid	x 6 (usually) [20]	
square centimeter	square feet	x 0.001 076 4	*x 0.001 1*
	square inches	x 0.155 00	*x 0.16*
	square meters	/10 000^ or x 1^E-04	
	square yards	x 1.196 0 E-04	*x 1.2 E-04*
square chain, surveyor's	acres	/10^ or x 0.1^	
	hectares	x 0.040 469	*x 0.04*
	square feet	x 4 356^	*x 4 400*
	square meters	x 404.69	*x 400*
	square miles	/6 400^ or x 1.562 5^E-04	*x 1.6 E-04*
	square rods	x 16^	
square foot	acres	x 2.295 7 E-05	*x 2.3 E-05*
	circular mils	x 1.833 5 E+08	*x 1.8 E+08*
	square centimeters	x 929.03	*x 900*
	square chains, surveyor's	x 2.295 7 E-04	*x 2.3 E-04*
	square inches	x 144^	*x 140*
	square meters	x 0.092 903	*x 0.09*
	square miles	x 3.587 0 E-08	*x 3.6 E-08*
	square rods	x 0.003 673 1	*x 0.003 7*
	square survey foot	x 1.000 0	*x 1*
	square yards	/9^ or x 0.111 11	*x 0.11*
	squares (builder's)	/100^	
square foot of heating surface	See "horsepower, boiler"		
square foot per second	centistokes	x 92 903	*x 90 000*
	square meters per second	x 0.092 903	*x 0.09*
	stokes	x 929.03	*x 900*
square foot (Can.)	square foot (US)	x 1^	
square foot, Paris (French land area, Queb.)	square centimeters	x 1 055.2	*x 1 100*
square foot, survey	square feet (intl.)	x 1.000 0	*x 1*
	square foot	x 1	
square hectometer	hectare	x 1^	
square inch	circular inches	x 4/pi or x 1.273 2	*x 1.3*
	circular mils	x 1.273 2 E+06	*x 1.3 E+06*
	square centimeters	x 6.451 6^	*x 6.5*
	square feet	/144^ or x 0.006 944 4	*x 0.007*
	square meters	x 6.451 6^E-04	*x 6.5 E-04*
	square millimeters	x 645.16^	*x 650*
	square yards	/1 296^ E-04 or x 7.716 0 E-04	*x 8 E-04*

[20] A "split" is half the usual volume of a beverage.

TABLE 2
Measurement Conversions, Alphabetical

All measurement units are US, unless otherwise noted. All number denominations "billion" and higher are US, unless otherwise noted.

CONVERT	TO EQUIVALENT	BY PRECISELY	OR WITHIN ± 5.0 %
square inch (Can.)	square inch (US)	x 1^	
square kilometer	acres	x 247.10	*x 250*
	square miles	x 0.386 10	*x 0.4*
square league (Texas)	acres	x 4 428.4	*x 4 400*
square meter	acres	x 2.471 0 E-04	*x 2.5 E-04*
	ares	/100^ or x 0.01^	
	circular mils	x 1.973 5 E+09	*x 2 E+09*
	hectares	/10 000^ or x 1^E-04	
	square centimeters	x 10 000^	
	square chains	x 0.002 471 1	*x 0.002 5*
	square feet	x 10.764	*x 11*
	square inches	x 1 550.0	*x 1 600*
	square miles	x 3.861 0 E-07	*x 4 E-07*
	square perches	x 0.039 537	*x 0.04*
	square poles	x 0.039 537	*x 0.04*
	square rods	x 0.039 537	*x 0.04*
	square yards	x 1.196 0	*x 1.2*
square micrometer	darcys	x 1.013 2	*x 1*
square mil	circular mils	x pi/4 or x 1.273 2	*x 1.3*
square mile	acres	x 640^	
	hectares	x 259.00	*x 260*
	quarter sections	x 4^	
	square centimeters	x 2.590 0 E+10	*x 2.6 E+10*
	square chains, surveyor's	x 6 400^	
	square feet	x 2.787 8 E+07	*x 2.8 E+07*
	square inches	x 4.014 5 E+09	*x 4 E+09*
	square kilometers	x 2.590 0	*x 2.6*
	square meters	x 2.590 0 E+06	*x 2.6 E+06*
	square rods	x 1.024^E+05	*x 1 E+05*
	square yards	x 3.097 6^ E+06	*x 3 E+06*
square mile (Can.)	square mile (US)	x 1^	
square mile (intl.)	square meters	x 2.590 0 E+06	*x 2.6 E+06*
	square mile, statute (US)	x 1^	
square mile, statute	square kilometers	x 2.590 0	*x 2.6*
	square mile	x 1^	
	square mile (intl.)	x 1^	
square millimeter	square inches	x 0.001 550 0	*x 0.001 6*
square millimeter per second	centistoke	x 1^	
square perch	square meters	x 25.293	*x 25*
	square rod	x 1^	
	square yards	x 30.25^	*x 30*
square pole	square meters	x 25.293	*x 25*
square rod	acres	/160^ or x 0.006 25^	*x 0.006*
	hectares	x 0.002 529 3	*x 0.002 5*
	square chains, surveyor's	/16^ or x 0.062 5^	*x 0.06*
	square feet	x 272.25^	*x 270*
	square meters	x 25.293	*x 25*
	square perch	x 1^	
	square yards	x 30.25^	*x 30*
square rod (Can.)	square rod (US)	x 1^	
square vara (Texas)	acres	/5 645 or x 1.771 5 E-04	*x 1.8 E-04*
square yard	acres	/4 840^ or x 2.066 1 E-04	*x 2 E-04*

TABLE 2
Measurement Conversions, Alphabetical

All measurement units are US, unless otherwise noted. All number denominations "billion" and higher are US, unless otherwise noted.

CONVERT	TO EQUIVALENT	BY PRECISELY	OR WITHIN ± 5.0 %
	ares	x 0.008 361 3	*x 0.008*
	circular mils	x 1.650 1 E+09	*x 1.7 E+09*
	square centimeters	x 8 361.3	*x 8 000*
	square feet	x 9^	
	square inches	x 1 296^	*x 1 300*
	square meters	x 0.836 13	*x 0.8*
	square perches	x 4/121^ or x 0.033 058	*x 0.033*
	square rods	x 4/121^ or x 0.033 058	*x 0.033*
square yard (Can.)	square yard (US)	x 1^	
square (builder's)	bundles	x 3^	
	square feet	x 100^	
standard distance (for capital ships)	yards	x 550	
standard distance (for cruisers)	yards	x 250	
statampere	amperes	x 3.335 6 E-10	*x 3.3 E-10*
statcoulomb	abcoulombs	x 3.335 6 E-11	*x 3.3 E-11*
	ampere-hours	x 9.265 7 E-14	*x 9 E-14*
	coulombs	x 3.335 6 E-10	*x 3.3 E-10*
	faradays (based on carbon-12)	x 3.457 1 E-15	*x 3.5 E-15*
statfarad	farads	x 1.112 7 E-12	*x 1.1 E-12*
stathenry	henrys	x 8.987 6 E+11	*x 9 E+11*
statmho	siemens	x 1.112 7 E-12	*x 1.1 E-12*
statohm	ohms	x 8.987 6 E+11	*x 9 E+11*
statvolt	volts	x 299.79	*x 300*
steam quality	steam percent (by weight) in a steam-water mixture	x 1^	
sterad	steradian	x 1^	
steradian	hemispheres	/2 pi or x 0.159 15	*x 0.16*
	spheres	/4 pi or x 0.079 577	*x 0.08*
	spherical degrees	x 180/pi or x 57.296	*x 60*
	spherical right angles	x 2/pi or x 0.636 62	*x 0.64*
stere	cubic meter	x 1^	
	kiloliter	x 1^	
steregon	spherical degrees	x 720^	
	steradians	x 4 pi or x 12.566	*x 13*
sterling silver	percent (by weight) of silver in alloy	x 92.5	*x 90*
stilb	candelas per square centimeter	x 1^	
stoke	square feet per second	x 0.001 076 4	*x 0.001 1*
	square meters per second	x 1^E-04	
stone (Brit.)	pounds, avoirdupois	x 14^	
stone (obsolete, Brit.)	pounds, avoirdupois	x 16	
subsonic	Mach less than 1	x 1	
	object speed less than sound speed	x 1	
sulung (Eng.)	hide (Eng.)	x 1	
supersonic	Mach 1 to 5	x 1	
	object speed greater than speed of sound	x 1	
svedberg	seconds	x 1 E-13	
swimming pool, junior olympic	meters	x 25^	
swimming pool, olympic	meters	x 50^	
-T-			
tablespoon, measuring	cup	/16^ or x 0.062 5^	*x 0.06*

TABLE 2
Measurement Conversions, Alphabetical

All measurement units are US, unless otherwise noted. All number denominations "billion" and higher are US, unless otherwise noted.

CONVERT	TO EQUIVALENT	BY PRECISELY	OR WITHIN ± 5.0 %
	drams, fluid	x 4^	
	milliliters	x 14.787	*x 15*
	ounces, fluid	/2^ or x 0.5^	
	pints, liquid	/32^ or x 0.031 25^	
	teaspoons	x 3^	
tablespoon, measuring (Brit.)	ounces, Imperial fluid (Brit.)	x 5/8^ or x 0.625^	*x 15*
tablespoon, measuring (Can.)	ounces, Imperial fluid (Can.)	/2^ or x 0.5^	
teaspoon, measuring	drams, fluid	x 4/3^ or x 1.333 33	*x 1.3*
	milliliters	x 4.928 9	
	ounces, fluid	/6^ or x 0.166 67	*x 0.67*
	tablespoons	/3^ or x 0.333 33	*x 0.33*
teaspoon, measuring (Brit.)	ounces, Imperial fluid (Brit.)	x 5/24^ or x 0.208 33	
teaspoon, measuring (Can.)	ounces, Imperial fluid (Can.)	/6^ or x 0.166 67	
terdiurnal	day	/3	
ternary	based on the number 3	x 1^	
tesla	gammas	x 1^E+09	
	gauss	x 1^E+04	
	weber per square meter	x 1^	
tex (for yarns)	grams per meter	/1 000^ or x 0.001^	
	milligram per meter	x 1^	
therm (AGA)	Btu (IT)	x 1.000 00 E+05	*x 100 000*
therm (EEC)	Btu (IT)	x 1.000 00 E+05	*x 100 000*
	calories (IT)	x 2.520 0 E+07	*x 2.5 E+07*
	joules	x 1.055 1 E+08	*x 1.1 E+08*
therm (US)	Btu (IT)	x 9.997 61 E+04	*x 100 000*
	calories (IT)	x 2.519 8 E+07	*x 2.5 E+07*
	joules	x 1.054 8 E+08	*x 1.1 E+08*
thermal insulance	x 1/thermal conductance	x 1^	
therme	therm	x 1^	
time constant	0.632 12 (= {e - 1}/e)	x 1	
time of day	See APPENDIX, "time"		
time scale, geologic	See APPENDIX, "time"		
time zone	degrees of longitude (per Greenwich standard)	x 15	
time, daylight-saving	time, standard	- 1^ hour	
time, decay	See "decay time"		
time, standard	time, daylight-saving	+ 1^ hour	
tithe	parts	/10^ or x 0.1^	
ton of material per day	pounds of material per hour	x 250/3^ or x 83.333	*x 80*
ton of refrigeration	Btu (IT) per hour	x 12 000	
	Btu (IT) per minute	x 200	
	the heat absorbed to melt one short ton of ice in 24 hours	x 1	
	watts	x 3 517	*x 3 500*
ton of refrigeration (Brit.)	Btu per minute	x 237.6	*x 240*
ton of water per day (@ 4 deg. C)	cubic feet per hour	x 1.334 9	*x 1.3*
	gallons per minute	x 0.166 43	*x 0.17*
ton (Brit.)	hundredweights (Brit.)	x 20^	
	pounds, avoirdupois	x 2 240^	*x 2 200*

TABLE 2
Measurement Conversions, Alphabetical

All measurement units are US, unless otherwise noted. All number denominations "billion" and higher are US, unless otherwise noted.

CONVERT	TO EQUIVALENT	BY PRECISELY	OR WITHIN ± 5.0 %
	stones (Brit.)	x 160^	
	tons, short (US)	x 1.12^	*x 1.1*
ton (customary, US)	ton, short (US)	x 1^	
ton (explosive energy of one ton of TNT)	joules	x 4.184 E+09 [6]	*x 4 E+09*
tonnage, cargo	number of long tons (of cargo)	x 1	
	number of metric tons (of cargo)	x 1	
tonnage, deadweight	number of long tons (of fuel, passengers, and cargo fully loaded)	x 1	
	number of metric tons (of fuel, passengers, and cargo fully loaded)	x 1	
tonnage, displacement	See "tonnage, cargo" and "tonnage, deadweight"		
tonnage, gross register (for ships)	tonnage, gross	x 1^	
tonnage, gross (for ships)	total register-ton capacity less nation-defined spaces	x 1	
tonnage, net register (for ships)	tonnage, net	x 1^	
tonnage, net (for ships)	gross tonnage less nation-defined spaces unavailable for cargo	x 1	
tonnage, register under-deck	number of register tons capacity under tonnage deck	x 1	
tonnage, ship	tonnage, gross (usually)	x 1	
tonnage, vessel	See a specific other "tonnage"		
tonne	ton, metric	x 1^	
tons, short per cubic yard	grams per milliliter	x 1.186 6	*x 1.2*
	kilograms per cubic meter	x 1 186.6	*x 1.2*
	pounds per cubic foot	x 74 000/999^ or x 74.074	*x 74*
	pounds per cubic inch	x 0.042 867	*x 0.042*
	pounds per gallon	x 9.902 3	*x 10*
ton, assay	grams	x 175/6^ or x 29.167	*x 30*
	milligrams	x 87 500/3^ or x 29 167^	*x 30 000*
ton, displacement (for ships)	cubic feet of fresh water	x 35.9	*x 36*
	cubic feet of sea water	x 35	
ton, English water (Brit.)	gallons (US)	x 270.91	*x 270*
	gallons, Imperial (Brit.)	x 224^	*x 220*
ton, freight	ton, measurement	x 1^	
ton, gross	hundredweights, gross	x 20^	
	kilograms	x 1 016.0	*x 1 000*
	ton, long	x 1^	
ton, long	hundredweights, long	x 20^	
	hundredweights, short	x 22.4^	*x 22*
	kilograms	x 1 016.0	*x 1 000*
	ounces, avoirdupois	x 35 840^	*x 36 000*
	pounds	x 2 240^	*x 2 200*
	tons, metric	x 1.016 0	*x 1*
	tons, short	x 1.12^	*x 1.1*
	ton, gross	x 1^	
ton, long per cubic yard	kilograms per cubic meter	x 1 328.9	*x 1 300*
ton, long (Can.)	ton, long (US)	x 1^	

[6] The ton of TNT energy is defined, not measured.

TABLE 2
Measurement Conversions, Alphabetical

All measurement units are US, unless otherwise noted. All number denominations "billion" and higher are US, unless otherwise noted.

CONVERT	TO EQUIVALENT	BY PRECISELY	OR WITHIN ± 5.0 %
ton, long (of fresh water displaced by ships)	foot, cubic	x 35.9	*x 36*
ton, long (of seawater displaced by ships)	foot, cubic	x 35	
ton, long (US)	ton (Brit.)	x 1^	
ton, measurement (for ships)	cubic feet	x 40^	
ton, metric	hundredweights, short	x 22.046	*x 22*
	kilograms	x 1 000^	
	ounces, avoirdupois	x 35 274	*x 35 000*
	pounds	x 2 204.6	*x 2 200*
	tons, long	x 0.984 21	*x 1*
	tons, short	x 1.102 3	*x 1.1*
ton, net	kilograms	x 907.18	*x 900*
	pounds	x 2 000^	
	ton, short	x 1^	
ton, register	cubic feet	x 100^	
	cubic meters	x 2.831 7	*x 2.8*
ton, shipper's	pounds	x 2 240^	*x 2 200*
ton, shipping	ton, measurement	x 1^	
	bushels	x 32.143	*x 32*
	cubic feet	x 40^	
ton, shipping (Brit.)	bushels (US)	x 33.750	*x 33*
	bushels, Imperial (Brit.)	x 32.701	*x 33*
	cubic feet	x 42^	
ton, shipping (US)	bushels, Imperial (Brit.)	x 31.144	*x 30*
ton, short	hundredweights, net	x 20^	
	hundredweights, short	x 20^	
	kilograms	x 907.18	*x 900*
	ounces, avoirdupois	x 32 000^	
	pounds	x 2 000^	
	tons, long	x 0.892 86	*x 0.9*
	tons, metric	x 0.907 18	*x 0.9*
	ton, net	x 1^	
ton, short per hour	kilograms per second	x 0.252 00	*x 0.25*
ton, short (Can.)	ton, short (US)	x 1^	
ton, short (force)	kilonewtons	x 8.896 4	*x 9*
	newtons	x 8 896.4	*x 9 000*
ton, water (Brit.)	gallons (US)	x 270.91	*x 270*
	gallons, Imperial (Brit.)	x 224^	*x 220*
tornado	See APPENDIX, "wind speeds"		
torr	atmospheres, standard	/760^ or x 0.001 315 8	*x 0.001 3*
	dynes per square centimeter	x 1 333.2	*x 1 300*
	millimeters of mercury [13]	x 1	
	pascals	x 133.32	*x 130*
	pounds (force) per square inch	x 0.019 337	*x 0.02*
township	sections	x 36^	
	square miles	x 36^	
township (Can.)	township (US)	x 1^	
tpi	tracks per inch	x 1^	
tps	transactions per second	x 1^	

[13] Unless otherwise noted, liquid-head conversions are based on: a pressure of one standard atmosphere; temperature for mercury = 0.0 deg. C = 32.0 deg. F; temperature for water = 4.0 deg. C = 39.2 deg. F.

TABLE 2
Measurement Conversions, Alphabetical

All measurement units are US, unless otherwise noted. All number denominations "billion" and higher are US, unless otherwise noted.

CONVERT	TO EQUIVALENT	BY PRECISELY	OR WITHIN ± 5.0 %
Tralles scale	ethanol percent (by volume) in water, in deg. Tralles	x 1^	
transfer ratio (for transformers)	ratio of number of turns in secondary winding to turns in primary winding	x 1^	
transmission unit (for electric power)	neper	x 1^	
transmittance	10 to power of negative density (optical)	x 1	
transonic	Mach approximately 1	x 1	
	object speed near speed of sound	x 1	
tredecillion (Brit.)	1^E+78	x 1^	
tredecillion (US)	1^E+42	x 1^	
triad	group of three	x 1^	
tricenary	based on the number 30	x 1^	
tricentennial	years	x 300^	
triennium	years	x 3^	
trilateral	sides	x 3^	
trillion (Brit.)	quintillion (US)	x 1^	
trillion (US)	1^E+12	x 1^	
trimester	months	x 3^	
	school years	/3^ or x 0.333 33	*x 0.3*
trinary	group of three	x 1^	
trio	group of three	x 1^	
tripartite	parts	x 3^	
triple	three times as large	x 1^	
triumvirate	group of three	x 1^	
troika	triumvirate	x 1^	
tropical wind scale	See APPENDIX, "wind speeds"		
tub (for herring, Can.)	herring barrel	/2 or x 0.5	
tun (volume)	gallons	x 252 (usually)	
	hogsheads	x 4	
tun (volume, for ships' water, Brit.)	gallons, Imperial (Brit.)	x 210^	
tunnage	tonnage	x 1^	
twain (mark twain)	fathoms	x 2^	
-U-			
ultrasound	See "range"		
undecillion (Brit.)	1^E+66	x 1^	
undecillion (US)	1^E+36	x 1^	
undecimal	based on the number 11	x 1^	
unit pole	webers	x 1.256 6 E-07	*x 1.3 E-07*
unit, astronomical	See "astronomical unit"		
U-value	thermal conductance	x 1^	
-V-			
vacuum	pressure below atmospheric pressure	x 1	
vacuum, coarse	torrs	x 1 to x 760	
vacuum, high	torrs	x 0.001 or less	
vacuum, low	pressure slightly below atmospheric pressure	x 1	
var	volt-ampere reactive	x 1^	
vara (Texas)	inches	x 33.33	*x 33*
vicenary	based on the number 20	x 1^	
vicennial	years	x 20^	
vigesimal	vicenary	x 1^	
vigintillion (Brit.)	1^E+120	x 1^	

TABLE 2
Measurement Conversions, Alphabetical

All measurement units are US, unless otherwise noted. All number denominations "billion" and higher are US, unless otherwise noted.

CONVERT	TO EQUIVALENT	BY PRECISELY	OR WITHIN ± 5.0 %
vigintillion (US)	1^E+63	x 1^	
virgate (Eng.)	hide (Eng.)	/4^ or x 0.25^	
viscosity, dynamic	pascal-second	x 1^	
viscosity, kinematic	square meter per second	x 1^	
volt	abvolts	x 1^E+08	
	EMU of electric potential	x 1^E+08	
	ESU of electric potential	x 0.003 335 6	*x 0.003 3*
	statvolts	x 0.003 335 6	*x 0.003 3*
volt per centimeter	volts per inch	x 2.54^	*x 2.5*
volt per inch	volts per centimeter	x 50/127^ or x 0.393 70	*x 0.4*
volt, absolute	volts, international	x 0.999 67	*x 1*
-W-			
wales per inch (for fabric, e.g., corduroy)	number of ribs or ridges per inch	x 1^	
watt	Btu (IT) per hour	x 3.412 1	*x 3.4*
	Btu (IT) per second	x 9.478 2 E-04	*x 9 E-04*
	ergs per second	x 1^E+07	
	foot-pounds (force) per hour	x 2 655.2	*x 2 700*
	foot-pounds (force) per minute	x 44.254	*x 44*
	foot-pounds (force) per second	x 0.737 56	*x 0.74*
	horsepower	x 0.001 341 0	*x 0.001 3*
	joule per second	x 1^	
watt per square centimeter	Btu (IT) per day-square foot	x 76 080	*x 76 000*
	Btu (IT) per hour-square foot	x 3 170.0	*x 3 200*
	calories (IT) per hour-square centimeter	x 859.85	*x 860*
	calories (IT) per second-square centimeter	x 0.238 85	*x 0.24*
	watts per square meter	x 10 000^	
watt per square centimeter-deg. C	Btu (IT) per day-square foot-deg. F	x 42 266	*x 42 000*
	Btu (IT) per hour-square foot-deg. F	x 1 761.1	*x 1 800*
	calories (IT) per hour-square centimeter-deg. C	x 859.85	*x 900*
	calories (IT) per second-square centimeter-deg. C	x 0.238 85	*x 0.24*
watt per square centimeter-deg. C/centimeter	Btu (IT) per day-square foot-deg. F/inch	x 16 640	*x 17 000*
	Btu (IT) per hour-square foot-deg. F/foot	x 57.779	*x 60*
	calories (IT) per hour-square centimeter-deg. C/centimeter	x 859.85	*x 900*
	calories (IT) per second-square centimeter-deg. C/centimeter	x 0.238 85	*x 0.24*
watt per square inch	watts per square meter	x 1 550.0	*x 1 600*
watt, absolute	watts, international	x 0.999 84	*x 1*
watt, intl.	watts, absolute	x 1.000 2	*x 1*
watt-hour	Btu (IT)	x 3.412 1	*x 3.4*
	foot-pounds (force)	x 2 655.2	*x 2 700*
	horsepower-hours	x 0.001 341 0	*x 0.001 3*
	joules	x 3 600^	
	kilocalories	x 0.859 85	*x 0.9*
	meter-kilograms (force)	x 367.10	*x 370*
watt-second	joules	x 1^	
wave number	1/meters	x 1^	
	1/wave length	x 1^	

TABLE 2
Measurement Conversions, Alphabetical

All measurement units are US, unless otherwise noted. All number denominations "billion" and higher are US, unless otherwise noted.

CONVERT	TO EQUIVALENT	BY PRECISELY	OR WITHIN ± 5.0 %
WC	water column	x 1^	
weber	maxwells	x 1^E+08	
	unit poles	x 7.957 7 E+06	*x 8 E+06*
weber per square centimeter	gauss	x 1^E-08	
	lines per square inch	x 6.451 6^	*x 6.5*
weber per square inch	gauss	x 1.550 0 E+07	*x 1.6 E+07*
	webers per square centimeter	x 0.155 00	*x 0.16*
weber per square meter	tesla	x 1^	
weight, basis (for paper)	weight of a ream of paper	x 1^	
wind	See APPENDIX, "wind speeds"		
Wobbe index (for flammable gas)	ratio of heat of combustion to specific gravity	x 1^	
-X-			
-Y-			
yard	centimeters	x 91.44^	*x 90*
	feet	x 3^	
	inches	x 36^	
	miles	/1 760^ or x 5.681 8 E-04	*x 5.7 E-04*
yard of ale	pints, fluid	x 2 or x 3	
yard of land	1/4 or 0.25 acre, 1 rod wide	x 1	
yard (before 1959)	meters	x 3 600/3 937^ or x 0.914 40	*x 0.9*
yard (British)	yard (US)	x 1^	
yard (Can.)	yard (US)	x 1^	
yard (since 1959)	meters	x 0.914 4^	*x 0.9*
year (customary)	months (customary)	x 12^	
	year, calendar	x 1^	
	year, mean solar	x 1^	
year, academic	year, school	x 1^	
year, anomalistic	365 days 6 hr. 13 min. 53.1 sec.	x 1	
	days (customary)	x 365.26	*x 365*
year, astronomical	365 days 5 hr. 48 min. 46 sec.	x 1	
year, banker's	days	x 360^ or x 365^	
year, bissextile	leap year	x 1^	
year, calendar	days (December 31 midnight to midnight 12^ months later)	x 365^ or x 366^	
year, civil	year	x 1^	
year, common	seconds	x 3.153 6^E+07	*x 3.2 E+07*
	weeks	x 52.143	*x 52*
year, common [19]	days	x 365^	
year, great	years	x 25 800	*x 26 000*
year, Gregorian	days	x 365.242 5	*x 365*
year, Julian	days	x 365.25	*x 365*
year, leap	seconds	x 3.162 2^E+07	*x 3.2 E+07*
year, leap [19]	days	x 366^	
year, legal	year	x 1^	
year, light	See "light-year"		
year, lunar	lunar months	x 12^	

[19] In the Gregorian calendar, every year whose number is divisible by four is a leap year except for centesimal years that are not exactly divisible by 400. The year 2000 is a leap year, 2100 is not a leap year.

TABLE 2
Measurement Conversions, Alphabetical

All measurement units are US, unless otherwise noted. All number denominations "billion" and higher are US, unless otherwise noted.

CONVERT	TO EQUIVALENT	BY PRECISELY	OR WITHIN ± 5.0 %
year, mean solar	365 days 5 hr. 48 min. 45.5 sec. mean solar time	x 1	
	days	x 365.242 5	*x 365*
	hours	x 8 765.8	*x 9 000*
	minutes	x 5.259 5 E+05	*x 5 E+05*
	seconds	x 3.155 7 E+07	*x 3.2 E+07*
year, Platonic	year, great	x 1^	
year, school	quarters	x 4^	
	semesters	x 2^	
	trimesters	x 3^	
	year, academic	x 1^	
year, sidereal	365 days 6 hr. 9 min. 9.5 sec. mean solar time	x 1	
	days, mean solar	x 365.26	*x 365*
	seconds, mean solar	x 3.155 8 E+07	*x 3.2 E+07*
year, tropical	365 days 5 hr. 48 min. 45.5 sec. mean solar time	x 1	
	seconds	x 3.155 7 E+07	*x 3.2 E+07*
	year, mean solar	x 1^	
-Z-			
zone time	standard time within a time zone	x 1^	

CHAPTER 8

CONVERSION TABLES

By Groups

TABLE 3
Measurement Conversions, By Group

All measurement units are US, unless otherwise noted. All number denominations "billion" and higher are US, unless otherwise noted.

CONVERT	TO EQUIVALENT	BY PRECISELY	OR WITHIN ± 5.0 %
ACCELERATION (see ANGULAR ACCELERATION and LINEAR ACCELERATION)			
ANGLE, PLANE			
acute angle	angle less than 90 degrees	x 1	
circle	degrees (angle)	x 360^	
	points (compass)	x 32^	
degree (angle)	grades	x 10/9^ or x 1.111 1	*x 1.1*
	mils (angle)	x 160/9^ or x 17.778	*x 18*
	minutes (angle)	x 60^	
	radians	x pi/180 or x 0.017 453	*x 0.017*
	revolutions	/360^ or x 0.002 777 8	*x 0.002 8*
	seconds (angle)	x 3 600^	
foot per 100 feet	grade (angle), percent	x 1^	
grad	See "grade"		
grade	degrees (angle)	x 0.9^	
	radians	x pi/200 or x 0.015 708	*x 0.016*
latitude	angular distance, in degrees, from a given point to a specified circle or reference plane [1]	x 1^	
longitude	angular distance, in degrees or time, from a point east or west to the prime meridian in Greenwich, England	x 1^	
mil (angle, for artillery)	degrees (angle)	x 9/160^ or x 0.056 25^	*x 0.056*
	minutes (angle)	x 27/8^ or x 3.375^	*x 3.4*
	radians	x pi/3 200 or x 9.817 5 E-04	*x 0.001*
	revolutions	/6 400^ or x 1.562 5^E-04	*x 1.6*
	seconds (angle)	x 202.5^	*x 200*
minute (angle)	degrees (angle)	/60^ or x 0.016 667	*x 0.017*
	radians	x pi/10 800 or x 2.908 9 E-04	*x 3 E-04*
	revolutions	/21 600^ or x 4.629 6 E-05	*x 4.6 E-05*
	seconds (angle)	x 60^	
normal	perpendicular	x 1^	
obtuse angle	angle greater than 90 degrees	x 1	
octant	circle	/8^ or x 0.125^	
perpendicular	at right angles	x 1^	
point (compass)	circle	/32^ or x 0.031 25^	
	degrees (angle)	x 11.25^	*x 11*
point,quarter (compass)	degrees (angle)	x 45/16^ or x 2.812 5^	*x 2.8*
quadrant (angle)	degrees (angle)	x 90^	
	minutes (angle)	x 5 400^	
	quarter circle	x 1^	
	radians	x 1.570 8	*x 1.6*
	revolutions	/4^ or x 0.25^	
	seconds (angle)	x 3.24^E+05	*x 3.2 E+05*
quarter circle	degrees (angle)	x 90^	
	quadrant	x 1^	
radian	57 deg. 17 min. 44.806 sec.	x 1	
	degrees (angle)	x 180/pi or x 57.296	*x 60*
	grades	x 63.662	*x 64*
	mils (angle)	x 1 018.6	*x 1 000*
	minutes (angle)	x 10 800/pi or x 3 437.7	*x 3 500*

1 "Latitude" example: A latitude of the Earth is an angular distance north or south along a meridian, measured in the range of 0 to 90 degrees from the equator.

TABLE 3
Measurement Conversions, By Group

All measurement units are US, unless otherwise noted. All number denominations "billion" and higher are US, unless otherwise noted.

CONVERT	TO EQUIVALENT	BY PRECISELY	OR WITHIN ± 5.0 %
	quadrant (angle)	x 0.636 62	*x 0.64*
	revolutions	/2 pi or x 0.159 15	*x 0.16*
	seconds (angle)	x 6.48 E+05/pi or x 2.062 6 E+05	*x 2 E+05*
revolution	degrees (angle)	x 360^	
	mils (angle)	x 6 400^	
	quadrants (angle)	x 4^	
	radians	x 2 pi or x 6.283 2	*x 6.3*
right angle	degrees (angle)	x 90^	
second (angle)	degrees (angle)	/3 600^ or x 2.777 8 E-04	*x 2.8 E-04*
	minutes (angle)	/60^ or x 0.016 667	*x 0.017*
	radians	x pi/6.48^E+05 or x 4.848 1 E-06	*x 5 E-06*
	revolutions	/1 296^E+06 or x 7.716 0 E-07	*x 8 E-07*
semicircle	degrees (angle)	x 180^	
sextant	circle	/6^ or x 0.166 67	
ANGLE, SOLID			
hemisphere	spheres	/2^ or x 0.5^	
	spherical right angles	x 4^	
	steradians	x 2 pi or x 6.283 2	*x 6.3*
spheradian	steradian	x 1^	
sphere	hemispheres	x 2^	
	spherical right angles	x 8^	
	steradians	x 4 pi or x 12.566	*x 13*
spherical degree	solid right angles	/90^ or x 0.011 111	*x 0.011*
spherical right angle	hemispheres	/4^ or x 0.25^	
	spheres	/8^ or x 0.125^	
	steradians	x pi/2 or x 1.570 8	*x 1.6*
spherical solid angle	steradians	x 4 pi or x 12.566	*x 13*
sterad	steradian	x 1^	
steradian	hemispheres	/2 pi or x 0.159 15	*x 0.16*
	spheres	/4 pi or x 0.079 577	*x 0.08*
	spherical degrees	x 180/pi or x 57.296	*x 60*
	spherical right angles	x 2/pi or x 0.636 62	*x 0.64*
steregon	spherical degrees	x 720^	
	steradians	x 4 pi or x 12.566	*x 13*
ANGULAR ACCELERATION			
radian per second per second	revolutions per minute per minute	x 572.96	*x 600*
	revolutions per minute per second	x 9.549 3	*x 10*
	revolutions per second per second	x 0.159 15	*x 0.16*
revolution per minute per minute	radians per second per second	x 0.001 745 3	*x 0.001 7*
	revolutions per second per second	/3 600^ or x 2.777 8 E-04	*x 2.8 E-04*
revolution per minute per second	radians per second per second	x 0.104 72	*x 0.1*
	revolutions per minute per minute	x 60^	
	revolutions per second per second	/60^ or x 0.016 667	
revolution per second per second	radians per second per second	x 6.283 2	*x 6*
	revolutions per minute per minute	x 3 600^	
ANGULAR DISTANCE (see ANGLE, PLANE and ANGLE, SOLID)			
ANGULAR SPEED			

TABLE 3
Measurement Conversions, By Group

All measurement units are US, unless otherwise noted. All number denominations "billion" and higher are US, unless otherwise noted.

CONVERT	TO EQUIVALENT	BY PRECISELY	OR WITHIN ± 5.0 %
degree (angle) per second	revolutions per minute	/6^ or x 0.166 67	*x 0.17*
	revolutions per second	/360^ or x 0.002 777 8	*x 0.002 8*
rad (angle) per second	radian per second	x 1^	
radian per second	degrees (angle) per second	x 57.296	*x 57*
	revolutions per minute	x 9.549 3	*x 10*
	revolutions per second	x 0.159 15	*x 0.16*
revolution per minute	degrees (angle) per second	x 6^	
	radians per second	x 0.104 72	*x 0.1*
	revolutions per second	/60^ or x 0.016 667	*x 0.017*
revolution per second	degrees (angle) per second	x 360^	
	radians per second	x 2 pi or x 6.283 2	*x 6*
	revolutions per minute	x 60^	
rpm	revolution per minute	x 1^	
AREA			
acre	ares	x 40.469	*x 40*
	forties	/40^ or x 0.025^	
	hectares	x 0.404 69	*x 0.4*
	quarter sections	/160^ or x 0.006 25^	*x 0.006*
	square chains, surveyor's	x 10^	
	square feet	x 43 560^	*x 43 000*
	square kilometers	x 0.004 046 9	*x 0.004*
	square links, surveyor's	x 1^E+05	
	square meters	x 4 046.9	*x 4 000*
	square miles	/640^ or x 0.001 562 5^	*x 0.001 6*
	square rods	x 160^	
	square yards	x 4 840^	*x 5 000*
acre (Can.)	acre (US)	x 1^	
agate line	See LENGTH		
are	acres	x 0.024 710	*x 0.025*
	square feet	x 1 076.4	*x 1 100*
	square meters	x 100^	
	square yards	x 119.60	*x 120*
arpent (French land area, Queb.)	square feet (French land measure, Queb.)	x 32 400^	*x 32 000*
barn	square meters	x 1^E-28	
bundle (builder's)	square feet	x 20^, x 25^, or x 100/3^ (per type of shingle)	
	square (builder's)	/3^ or x 0.333 33	*x 0.33*
carucate (Eng.)	hide	x 1	
centare	square meter	x 1^	
centiare	centare	x 1^	
circular inch	circular mils	x 1^E+06	
	square inches	x pi/4 or x 0.785 40	*x 0.8*
circular mil	circular inches	x 1^E-06	
	square feet	x 5.454 2 E-09	*x 5.5 E-09*
	square inches	x 7.854 0 E-07	*x 8 E-07*
	square meters	x 5.067 1 E-10	*x 5 E-10*
	square mils	x pi/4 or x 0.785 40	*x 0.8*
	square yards	x 6.060 2 E-10	*x 6 E-10*
foot, superficial	square foot	x 1^	
forty	acres	x 40^	
	quarter sections	/4^ or x 0.25^	
hectare	acres	x 2.471 0	*x 2.5*

TABLE 3
Measurement Conversions, By Group

All measurement units are US, unless otherwise noted. All number denominations "billion" and higher are US, unless otherwise noted.

CONVERT	TO EQUIVALENT	BY PRECISELY	OR WITHIN ± 5.0 %
	square chains, surveyor's	x 24.710	*x 25*
	square feet	x 1.076 4 E+05	*x 1.1 E+05*
	square hectometer	x 1^	
	square meters	x 10 000^	
	square miles	x 0.003 861 0	*x 3/800 or x 0.004*
	square rods	x 395.37	*x 400*
hide (Eng.)	acres	x 120	
inch, superficial	square inch	x 1^	
labor (Texas)	acres	x 177.1	*x 180*
league	Also see LENGTH		
league (Texas)	acres	x 4 428.4	*x 4 400*
perch	Also see LENGTH and VOLUME		
perch (for land, Brit.)	square rod	x 1^	
perch (French land area, Queb.)	square feet (French land measure, Queb.)	x 324^	*x 320*
quad (area)	quadrangle	x 1^	
quarter section	acres	x 160^	
	forties	x 4^	
rod	Also see LENGTH and VOLUME		
	square yards	x 30.25^	*x 30*
rood (area, Brit.)	acres	/4^ or x 0.25^	
	Also see LENGTH		
	square rods	x 40^	
	square yards	x 1 210^	*x 1 200*
rood (Can.)	hectares	x 0.101 17	*x 0.1*
section	square mile	x 1^	
specific surface	ratio of surface area of a powder to its mass	x 1^	
	ratio of total surface of adsorbent substance to its volume	x 1^	
square centimeter	square feet	x 0.001 076 4	*x 0.001 1*
	square inches	x 0.155 00	*x 0.16*
	square meters	/10 000^ or x 1^E-04	
	square yards	x 1.196 0 E-04	*x 1.2 E-04*
square chain, surveyor's	acres	/10^ or x 0.1^	
	hectares	x 0.040 469	*x 0.04*
	square feet	x 4 356^	*x 4 400*
	square meters	x 404.69	*x 400*
	square miles	/6 400^ or x 1.562 5^E-04	*x 1.6 E-04*
	square rods	x 16^	
square foot	acres	x 2.295 7 E-05	*x 2.3 E-05*
	circular mils	x 1.833 5 E+08	*x 1.8 E+08*
	square centimeters	x 929.03	*x 900*
	square chains, surveyor's	x 2.295 7 E-04	*x 2.3 E-04*
	square inches	x 144^	*x 140*
	square meters	x 0.092 903	*x 0.09*
	square miles	x 3.587 0 E-08	*x 3.6 E-08*
	square rods	x 0.003 673 1	*x 0.003 7*
	square survey foot	x 1.000 0	*x 1*
	square yards	/9^ or x 0.111 11	*x 0.11*
	squares (builder's)	/100^	
square foot of heating surface	See POWER, "horsepower, boiler"		

TABLE 3
Measurement Conversions, By Group

All measurement units are US, unless otherwise noted. All number denominations "billion" and higher are US, unless otherwise noted.

CONVERT	TO EQUIVALENT	BY PRECISELY	OR WITHIN ± 5.0 %
square foot (Can.)	square foot (US)	x 1^	
square foot, Paris (French land area, Queb.)	square centimeters	x 1 055.2	*x 1 100*
square foot, survey	square feet (intl.)	x 1	
	square foot	x 1	
square hectometer	hectare	x 1^	
square inch	circular inches	x 4/pi or x 1.273 2	*x 1.3*
	circular mils	x 1.273 2 E+06	*x 1.3 E+06*
	square centimeters	x 6.451 6^	*x 6.5*
	square feet	/144^ or x 0.006 944 4	*x 0.007*
	square meters	x 6.451 6^E-04	*x 6.5 E-04*
	square millimeters	x 645.16^	*x 650*
	square yards	/1 296^ E-04 or x 7.716 0 E-04	*x 8 E-04*
square inch (Can.)	square inch (US)	x 1^	
square kilometer	acres	x 247.10	*x 250*
	square miles	x 0.386 10	*x 0.4*
square league (Texas)	acres	x 4 428.4	*x 4 400*
square meter	acres	x 2.471 0 E-04	*x 2.5 E-04*
	ares	/100^ or x 0.01^	
	circular mils	x 1.973 5 E+09	*x 2 E+09*
	hectares	/10 000^ or x 1^E-04	
	square centimeters	x 10 000^	
	square chains	x 0.002 471 1	*x 0.002 5*
	square feet	x 10.764	*x 11*
	square inches	x 1 550.0	*x 1 600*
	square miles	x 3.861 0 E-07	*x 4 E-07*
	square perches	x 0.039 537	*x 0.04*
	square poles	x 0.039 537	*x 0.04*
	square rods	x 0.039 537	*x 0.04*
	square yards	x 1.196 0	*x 1.2*
square micrometer	darcys	x 1.013 2	*x 1*
square mil	circular mils	x pi/4 or x 1.273 2	*x 1.3*
square mile	acres	x 640^	
	hectares	x 259.00	*x 260*
	quarter sections	x 4^	
	square centimeters	x 2.590 0 E+10	*x 2.6 E+10*
	square chains, surveyor's	x 6 400^	
	square feet	x 2.787 8 E+07	*x 2.8 E+07*
	square inches	x 4.014 5 E+09	*x 4 E+09*
	square kilometers	x 2.590 0	*x 2.6*
	square meters	x 2.590 0 E+06	*x 2.6 E+06*
	square rods	x 1.024^E+05	*x 1 E+05*
	square yards	x 3.097 6^ E+06	*x 3 E+06*
square mile (Can.)	square mile (US)	x 1^	
square mile (intl.)	square meters	x 2.590 0 E+06	*x 2.6 E+06*
	square mile, statute (US)	x 1^	
square mile, statute	square kilometers	x 2.590 0	*x 2.6*
	square mile	x 1^	
	square mile (intl.)	x 1^	
square millimeter	square inches	x 0.001 550 0	*x 0.001 6*
square perch	square meters	x 25.293	*x 25*
	square rod	x 1^	

TABLE 3
Measurement Conversions, By Group

All measurement units are US, unless otherwise noted. All number denominations "billion" and higher are US, unless otherwise noted.

CONVERT	TO EQUIVALENT	BY PRECISELY	OR WITHIN ± 5.0 %
	square yards	x 30.25^	*x 30*
square pole	square meters	x 25.293	*x 25*
square rod	acres	/160^ or x 0.006 25^	*x 0.006*
	hectares	x 0.002 529 3	*x 0.002 5*
	square chains, surveyor's	/16^ or x 0.062 5^	*x 0.06*
	square feet	x 272.25^	*x 270*
	square meters	x 25.293	*x 25*
	square perch	x 1^	
	square yards	x 30.25^	*x 30*
square rod (Can.)	square rod (US)	x 1^	
square vara (Texas)	acres	/5 645 or x 1.771 5 E-04	*x 1.8 E-04*
square yard	acres	/4 840^ or x 2.066 1 E-04	*x 2 E-04*
	ares	x 0.008 361 3	*x 0.008*
	circular mils	x 1.650 1 E+09	*x 1.7 E+09*
	square centimeters	x 8 361.3	*x 8 000*
	square feet	x 9^	
	square inches	x 1 296^	*x 1 300*
	square meters	x 0.836 13	*x 0.8*
	square perches	x 4/121^ or x 0.033 058	*x 0.033*
	square rods	x 4/121^ or x 0.033 058	*x 0.033*
square yard (Can.)	square yard (US)	x 1^	
square (builder's)	bundles	x 3^	
	square feet	x 100^	
sulung (Eng.)	hide (English)	x 1	
township	sections	x 36^	
	square miles	x 36^	
township (Can.)	township (US)	x 1^	
virgate (Eng.)	hide (Eng.)	/4^ or x 0.25^	
yard of land	1/4 or 0.25 acre, 1 rod wide	x 1	
BENDING MOMENT (see TORQUE)			
BENDING MOMENT PER UNIT LENGTH (see TORQUE PER UNIT LENGTH)			
CAPACITY (see VOLUME)			
COMPOSITION (see DENSITY)			
COMPUTER			
baud	bit per second	x 1^	
	characters per second	/8^ or x 0.125^	*x 0.13*
ber	bit error rate	x 1^	
bips	billion (US) instructions per second	x 1^	
bit	binary digit	x 1^	
	bytes	/8^ or x 0.125^	*x 0.13*
	nibbles	/4^ or x 0.25^	
bpi	bits per inch	x 1^	
bps	bits per second	x 1^	
byte	bits	x 8^	
	nibbles	/2^ or x 0.5^	
cpi	characters per inch	x 1^	
dpi	dots per inch	x 1^	
floating-point operations per second	calculations per second	x 1^	

TABLE 3
Measurement Conversions, By Group

All measurement units are US, unless otherwise noted. All number denominations "billion" and higher are US, unless otherwise noted.

CONVERT	TO EQUIVALENT	BY PRECISELY	OR WITHIN ± 5.0 %
flop	floating-point operation	x 1^	
flops	floating-point operations per second	x 1^	
GB	gigabyte	x 1^	
Gflops	gigaflops	x 1^	
gigabyte	bytes	x 2^ to power 30 or x 1.073 7 E+09	*x billion*
gigaflops	billion (US) floating-point operations per second	x 1	
ips	instructions per second	x 1^	
K	kilobyte	x 1^	
kilobaud	bauds	x 2^ to power 10 or x 1 024^	*x thousand*
kilobit	bits	x 2^ to power 10 or x 1 024^	*x thousand*
kilobyte	bytes	x 2^ to power 10 or x 1 024^	*x thousand*
MB	megabyte	x 1^	
megabyte	bytes	x 2^ to power 20 or x 1.048 6 E+06	*x million*
megaflops	million floating-point operations per second	x 1^	
mips	milion instructions per second	x 1^	
M-flops	megaflops	x 1^	
nibble	bits	x 4^	
	bytes	x 2^	
nybble	nibble	x 1^	
ppm (for computers)	pages (8.5 x 11 inches, double-spaced) per minute	x 1^	
tpi	tracks per inch	x 1^	
tps	transactions per second	x 1^	
CONCENTRATION (see DENSITY)			
DENSITY; CONCENTRATION; and COMPOSITION			
absolute alcohol	ethanol-water solution containing 5.1 percent , or less, water by volume	x 1^	
Balling scale (for brewing industry)	dissolved-solids weight percentages, in deg. Balling	x 1^	
bone dry	percent moisture	zero (nominal)	
Brix scale (for sugar industry)	sucrose weight percentages in water solution, in deg. Brix	x 1^	
cetane number (diesel-fuel ignition rating)	volume percent of cetane in a standard reference fuel	x 1^	
cetane rating	cetane number	x 1^	
concentration, molal	moles of solute per kilogram of solvent	x 1^	
concentration, molar	moles of solute per liter of solution	x 1^	
concentration, normal	gram-equivalent of solute per liter of solution	x 1^	
concentration, percent (by mass)	grams of solute per 100 grams of solution	x 1^	
concentration, percent (by volume)	milliliters of solute per 100 milliliters of solution	x 1^	
concentration, relative	moles per cubic meter	x 1^	
density	1/specific volume	x 1^	
density of water (@ 104 deg. F)	pounds per cubic foot	x 61.942	*x 60*
	pounds per gallon	x 8.280 4	*x 8*
density of water (@ 140 deg. F)	pounds per cubic foot	x 61.390	*x 60*
	pounds per gallon	x 8.206 7	*x 8*
density of water (@ 4 deg. C)	pounds per cubic foot	x 62.426	*x 60*
	pounds per gallon	x 8.345 2	*x 8*

TABLE 3
Measurement Conversions, By Group

All measurement units are US, unless otherwise noted. All number denominations "billion" and higher are US, unless otherwise noted.

CONVERT	TO EQUIVALENT	BY PRECISELY	OR WITHIN ± 5.0 %
density of water (@ 68 deg. F)	pounds per cubic foot	x 62.316	*x 60*
	pounds per gallon	x 8.330 4	*x 8*
density, relative	specific gravity	x 1^	
fineness (for precious-metal alloys)	parts per thousand (by weight)	x 1^	
fineness (of gold or silver)	parts per thousand (by weight)	x 1^	
grain per cubic foot	milligrams per cubic meter	x 2 288.4	*x 2 200*
	pounds per thousand cubic feet	/7^ or x 0.142 86	*x 0.14*
grain per gallon	pounds per gallon	/7 000^ or x 1.428 6 E-04	*x 1.4 E-04*
	grains per Imperial gallon (Brit.)	x 1.201 0	*x 1.2*
	grams per cubic meter	x 17.118	*x 17*
	kilograms per cubic meter	x 0.017 118	*x 0.017*
	pounds per million gallons	x 1 000/7^ or x 142.86	*x 140*
grain per gallon (of water @ 20 deg. C)	parts per million (by weight)	x 17.149	*x 17*
grain per gallon (of water @ 4 deg. C)	parts per million (by weight)	x 17.118	*x 17*
grain per Imperial gallon (Brit.)	pounds per Imperial gallon (Brit.)	/7 000^ or x 1.428 6 E-04	*x 1.4 E-04*
	pounds per million Imperial gallons (Brit.)	x 1 000/7^ or x 142.86	*x 140*
grain per Imperial gallon (of water @ 20 deg. C, Brit.)	grains per gallon (US)	x 0.832 67	*x 0.8*
	parts per million (by weight)	x 14.279	*x 14*
grain per Imperial gallon (of water @ 4 deg. C, Brit.)	parts per million (by weight)	x 14.254	*x 14*
gram per cubic centimeter	pounds per cubic foot	x 62.428	*x 60*
	pounds per cubic inch	x 0.036 127	*x 0.036*
gram per cubic meter	grains per cubic foot	x 0.437 00	*x 4/9 or x 0.44*
	grains per gallon	x 0.058 418	*x 0.06*
	milligram per liter	x 1^	
gram per liter	grains per gallon	x 58.418	*x 60*
	kilogram per cubic meter	x 1^	
	pounds per 1 000 gallons	x 8.345 4	*x 8*
	pounds per cubic foot	x 0.062 428	*x 0.06*
gram per liter (of water @ 4 deg. C)	parts per million (by weight)	x 1 000^	
gram per milliliter	kilograms per cubic meter	x 1 000^	
	pounds per cubic foot	x 62.428	*x 60*
	pounds per cubic inch	x 0.036 127	*x 0.036*
	pounds per gallon	x 8.345 4	*x 8*
	tons, short per cubic yard	x 0.842 78	*x 0.84*
humidity, molal	mols of water per mol of dry gas	x 1^	
	mols of water per mols of dry gas	x 1^	
humidity, specific	mass of water vapor per unit mass of moist air	x 1^	
karat (of gold) [2]	grams of gold per kilogram of alloy	x 125/3^ or x 41.667	*x 40*
	percent by weight of gold in alloy	x 25/6^ or x 4.166 7	*x 4*
kilogram per cubic meter	grains per gallon	x 58.418	*x 60*
	gram per liter	x 1^	
	grams per milliliter	/1 000^ or x 0.001^	

[2] 100-percent pure gold has 24 karats of gold.

TABLE 3
Measurement Conversions, By Group

All measurement units are US, unless otherwise noted. All number denominations "billion" and higher are US, unless otherwise noted.

CONVERT	TO EQUIVALENT	BY PRECISELY	OR WITHIN ± 5.0 %
	ounces per cubic inch	x 5.780 4 E-04	*x 6 E-04*
	ounces per gallon	x 0.133 53	*x 0.13*
	ounces per gallon, Imperial (Brit.)	x 0.160 36	*x 0.16*
	pounds per cubic foot	x 0.06 242 8	*x 0.06*
milligram per liter (of water @ 4 deg. C)	parts per million (by weight)	x 1^	
molal solution	See "concentration, molal"		
molality	molal solution	x 1^	
molar solution	See "concentration, molar"		
molarity	molar solution	x 1^	
mole fraction	moles of solute per mole of solution	x 1^	
mole percent	mole fraction	/100^ or x 0.01^	
	moles of solute per 100 moles of solution	x 1^	
normal solution	See "concentration, normal"		
octane number (gasoline anti-knock rating)	volume percent of isooctane in a standard reference fuel	x 1^	
octane rating	octane number	x 1^	
ounce per cubic inch	kilograms per cubic meter	x 1 730.0	*x 1 700*
ounce per gallon	kilograms per cubic meter	x 7.489 2	*x 7.5*
ounce per Imperial gallon (Brit.)	kilograms per cubic meter	x 6.236 0	*x 6*
part per million (by weight)	grains per gallon (of water @ 20 deg. C)	x 0.058 313	*x 0.06*
	grains per gallon (of water @ 4 deg. C)	x 0.058 417	*x 0.06*
	grains per Imperial gallon (of water @ 20 deg. C, Brit.)	x 0.070 032	*x 0.07*
	grains per Imperial gallon (of water @ 4 deg. C, Brit.)	x 0.070 155	*x 0.07*
	grams per liter (of water @ 4 deg. C)	x 0.001^	
	pounds per gallon (of water @ 20 deg. C)	x 8.330 5 E-06	*x 8 E-06*
	pounds per gallon (of water @ 4 deg. C)	x 8.345 2 E-06	*x 8 E-06*
	pounds per Imperial gallon (of water @ 20 deg. C, Brit.)	x 1.000 5 E-05	*x 1 E-05*
	pounds per Imperial gallon (of water @ 4 deg. C, Brit.)	x 1.002 2 E-05	*x 1 E-05*
	pounds per million gallons (of water @ 20 deg. C)	x 8.330 5	*x 8*
	pounds per million gallons (of water @ 4 deg. C)	x 8.345 2	*x 8*
	pounds per million Imperial gallons (of water @ 20 deg. C, Brit.)	x 10.005	*x 10*
	pounds per million Imperial gallons (of water @ 4 deg. C, Brit.)	x 10.022	*x 10*
percent of gold in alloy (by weight)	karats [2]	x 6/25^ or x 0.24^	
	parts per million (by weight)	x 10 000	
percent solution (by mass)	grams of solute per 100 grams of solution	x 1^	
percent solution (by volume) [3]	liters of solute per 100 liters of solution	x 1^	
pH (= power of hydrogen ion)	negative log of hydrogen-ion activity in gram-ions per liter	x 1^	
pIon (= power of any specified ion)	negative log of specified-ion activity in gram-ions per liter	x 1^	
pound per cubic foot	grams per milliliter	x 0.016 018	*x 0.016*
	kilograms per cubic meter	x 16.018	*x 16*

2 100-percent pure gold has 24 karats of gold.

3 The volume of a solution may differ from the sum of the separate volumes of the solute and solvent.

TABLE 3
Measurement Conversions, By Group

All measurement units are US, unless otherwise noted. All number denominations "billion" and higher are US, unless otherwise noted.

CONVERT	TO EQUIVALENT	BY PRECISELY	OR WITHIN ± 5.0 %
	pounds per cubic inch	x 5.787 0 E-04	*x 6 E-04*
	pounds per gallon	x 0.133 68	*x 0.13*
	tons, short per cubic yard	x 0.013 5^	*x 0.013*
pound per cubic inch	grams per milliliter	x 27.680	*x 28*
	kilograms per cubic meter	x 27 680	*x 28 000*
	pounds per cubic foot	x 1 728^	*x 1 700*
	pounds per gallon	x 231^	*x 230*
	tons, short per cubic yard	x 23.328^	*x 23*
pound per cubic yard	kilograms per cubic meter	x 0.593 28	*x 0.6*
pound per gallon	grains per cubic inch	x 1 000/33^ or x 30.303	*x 30*
	grams per liter	x 0.119 83	*x 0.12*
	kilograms per cubic meter	x 119.83	*x 120*
	ounces per cubic inch	x 0.069 264	*x 0.07*
	pounds per cubic foot	x 7.480 5	*x 7.5*
	tons, short per cubic yard	x 0.100 99	*x 0.1*
	grains per gallon	x 7 000^	
	pounds per million gallons	x 1^E+06	
pound per gallon (of water @ 20 deg. C)	parts per million (by weight)	x 1.200 4 E+05	*x 1.2 E+05*
pound per gallon (of water @ 4 deg. C)	parts per million (by weight)	x 1.198 3 E+05	*x 1.2 E+05*
pound per Imperial gallon (Brit.)	grains per Imperial gallon	x 7 000^	
	kilograms per cubic meter	x 99.776	*x 100*
pound per Imperial gallon (of water @ 20 deg. C)	parts per million (by weight)	x 99 955	*x 1 E+05*
pound per Imperial gallon (of water @ 4 deg. C)	parts per million (by weight)	x 99 780	*x 1 E+05*
pound per million gallons	grains per gallon	x 0.007^	
	pounds per gallon	x 1^E-06	
pound per million gallons (of water @ 20 deg. C)	parts per million (by weight)	x 0.120 04	*x 0.12*
pound per million gallons (of water @ 4 deg. C)	parts per million (by weight)	x 0.119 83	*x 0.12*
pound per million Imperial gallons (Brit.)	grains per Imperial gallon (Brit.)	x 0.007^	
pound per million Imperial gallons (of water @ 20 deg. C, Brit.)	parts per million (by weight)	x 0.099 955	*x 0.1*
pound per million Imperial gallons (of water @ 4 deg. C, Brit.)	parts per million (by weight)	x 0.099 779	*x 0.1*
ppb	parts per billion	x 1^	
ppm (concentration)	parts per million	x 1^	
proof spirit (Brit.)	57.10 percent ethanol (by volume) in distilled water	x 1^	
proof spirit (U.S. standard concentration)	50 percent ethanol (by volume) in water (@ 60 deg. F)	x 1^	
proof (for liquors)	ethanol percent (by volume) in water	x 50	
relative density	See "density, relative"		
Richter scale (concentration)	ethanol percent (by volume) in water, in deg. Richter	x 1^	

TABLE 3
Measurement Conversions, By Group

All measurement units are US, unless otherwise noted. All number denominations "billion" and higher are US, unless otherwise noted.

CONVERT	TO EQUIVALENT	BY PRECISELY	OR WITHIN ± 5.0 %
Sikes scale	ethanol percent (by volume) in water, in deg. Sikes	x 1^	
slug per cubic foot	kilograms per cubic meter	x 515.38	*x 500*
steam quality	steam percent (by weight) in a steam-water mixture	x 1^	
sterling silver	percent (by weight) of silver in alloy	x 92.5	*x 90*
tons, short per cubic yard	grams per milliliter	x 1.186 6	*x 1.2*
	kilograms per cubic meter	x 1 186.6	*x 1.2*
	pounds per cubic foot	x 74 000/999^ or x 74.074	*x 74*
	pounds per cubic inch	x 0.042 867	*x 0.042*
	pounds per gallon	x 9.902 3	*x 10*
ton, long per cubic yard	kilograms per cubic meter	x 1 328.9	*x 1 300*
Tralles scale	ethanol percent (by volume) in water, in deg. Tralles	x 1^	
DISTANCE (see ANGLE, PLANE; ANGLE, SOLID; and LENGTH)			
ELECTRICITY			
abampere	amperes	x 10^	
	coulombs per second	x 0.1^	
	faradays (based on carbon-12) per second	x 1.036 4 E-04	*x 1 E-04*
	faradays (chemical) per second	x 1.036 3 E-04	*x 1 E-04*
	faradays (physical) per second	x 1.036 0 E-04	*x 1 E-04*
	statamperes	x 2.997 9 E+10	*x 3 E+10*
abcoulomb	ampere-hours	/360^ or x 0.002 777 8	*x 0.002 8*
	coulombs	x 10^	
	faradays (based on carbon-12)	x 1.036 4 E-04	*x 1 E-04*
	statcoulombs	x 2.997 9 E+10	*x 3 E+10*
abfarad	farads	x 1^E+09	
abhenry	henrys	x 1^E-09	
abmho	mhos	x 1^E+09	
	siemens	x 1^E+09	
abohm	ohms	x 1^E-09	
abohm per centimeter cube	abohm-centimeter	x 1^	
abvolt	volts	x 1^E-08	
admittance	1/impedance	x 1^	
ampere	abamperes	/10^ or x 0.1^	
	coulomb per second	x 1^	
	EMU of current	/10^ or x 0.1^	
	ESU of current	x 2.997 9 E+09	*x 3 E+09*
	gilberts	x 1.256 6	*x 1.3*
	statamperes	x 2.997 9 E+09	*x 3 E+09*
ampere per meter	oersteds	x 0.012 566	*x 0.013*
ampere per square centimeter	amperes per square inch	x 6.451 6^	*x 6.5*
	amperes per square meter	x 1^E+04	
ampere per square inch	amperes per square centimeter	x 0.155 00	*x 0.16*
	amperes per square meter	x 1 550.0	*x 1 500*
ampere per square meter	amperes per square centimeter	x 1^E-04	
	amperes per square inch	x 6.451 6^E-04	*x 6.5 E-04*
ampere, absolute	amperes, international	x 1.000 2	*x 1*
ampere-hour	abcoulombs	x 360^	

TABLE 3
Measurement Conversions, By Group

All measurement units are US, unless otherwise noted. All number denominations "billion" and higher are US, unless otherwise noted.

CONVERT	TO EQUIVALENT	BY PRECISELY	OR WITHIN ± 5.0 %
	coulombs	x 3 600^	
	faradays (based on carbon-12)	x 0.037 311	*x 0.037*
	faradays, chemical	x 0.037 307	*x 0.037*
	faradays, physical	x 0.037 297	*x 0.037*
	statcoulombs	x 1.079 3 E+13	*x 1.1 E+13*
biot	amperes	x 10^	
coulomb	abcoulombs	/10^ or x 0.1^	
	ampere-hours	/3 600^ or x 2.777 8 E-04	*x 2.8 E-04*
	ampere-second	x 1^	
	faradays (based on carbon-12)	x 1.036 4 E-05	*x 1 E-05*
	faradays, chemical	x 1.036 3 E-05	*x 1 E-05*
	faradays, physical	x 1.036 0 E-05	*x 1 E-05*
	statcoulombs	x 2.997 9 E+09	*x 3 E+09*
coulomb per square centimeter	coulombs per square inch	x 6.451 6^	*x 6.5*
coulomb per square inch	coulombs per square centimeter	x 0.155 00	*x 0.16*
coulomb, absolute	coulombs, international	x 1.000 2	*x 1*
current density	ampere per square meter	x 1^	
decibel (for current)	ratio of two levels of current with equal resistance	x 10 to power {no. of decibels x 0.05}	
decibel (for voltage)	ratio of two levels of voltage with equal resistance	x 10 to power {no. of decibels x 0.05}	
electric charge density	coulomb per cubic meter	x 1^	
electric field strength	volt per meter	x 1^	
electric flux density	coulomb per square meter	x 1^	
electromagnetic spectrum	See APPENDIX, "electromagnetic spectrum"		
EMU of capacitance	farads	x 1^E+09	
EMU of current	amperes	x 10^	
EMU of electric potential	volts	x 1^E-08	
EMU of inductance	henrys	x 1^E-09	
EMU of resistance	ohms	x 1^E-09	
ESU of capacitance	farads	x 1.112 7 E-12	*x 1.1 E-12*
ESU of current	amperes	x 3.335 6 E-10	*x 3.3 E-10*
ESU of electric potential	volts	x 299.79	*x 300*
ESU of inductance	henrys	x 8.987 6 E+11	*x 9 E+11*
ESU of resistance	ohms	x 8.987 6 E+11	*x 9 E+11*
farad	abfarads	x 1^E-09	
	EMU of capacitance	x 1^E-09	
	ESU of capacitance	x 8.987 6 E+11	*x 9 E+11*
	statfarads	x 8.987 6 E+11	*x 9 E+11*
faraday (based on carbon-12)	abcoulombs	x 9 648.7	*x 10 000*
	ampere-hours	x 26.302	*x 26*
	coulombs	x 96 487	*x 100 000*
	statcoulombs	x 2.892 6 E+14	*x 3 E+14*
faraday, chemical	coulombs	x 96 496	*x 100 000*
faraday, physical	ampere-hours	x 26.812	*x 27*
	coulombs	x 96 522	*x 100 000*
farad, absolute	farads, international	x 1.000 5	*x 1*
henry	abhenrys	x 1^E+09	
	EMU of inductance	x 1^E+09	
	ESU of inductance	x 1.112 6 E-12	*x 1.1 E-12*

TABLE 3
Measurement Conversions, By Group

All measurement units are US, unless otherwise noted. All number denominations "billion" and higher are US, unless otherwise noted.

CONVERT	TO EQUIVALENT	BY PRECISELY	OR WITHIN ± 5.0 %
	stathenrys	x 1.112 6 E-12	*x 1.1 E-12*
henry, absolute	henrys, international	x 0.999 51	*x 1*
impedance	1/admittance	x 1^	
joule, absolute	joules, international	x 0.999 84	*x 1*
mho	1/ohm	x 1^	
	siemens	x 1^	
microhm per centimeter cube	microhm-centimeter	x 1^	
microhm per inch cube	microhm-inch	x 1^	
microhm-centimeter	microhm-inches	x 50/127^ or x 0.393 70	*x 0.4*
	ohms (mil, foot)	x 6.015 3	*x 6*
	ohm-meters	x 1^E-08	
microhm-inch	microhm-centimeters	x 2.54^	*x 2.5*
	ohms (mil, foot)	x 15.279	*x 15*
	ohm-meters	x 2.54^E-08	*x 2.5 E-08*
neper	decibels	x 8.685 9	*x 8.7*
	ratio of two power levels	x e or 2.718 3 to power of {no. of nepers x 2}	
ohm	1/mho	x 1^	
	abohms	x 1^E+09	
	EMU of resistance	x 1^E+09	
	ESU of resistance	x 1.112 6 E-12	*x 1.1 E-12*
	statohms	x 1.112 6 E-12	*x 1.1 E-12*
ohm per mil-foot	ohm (mil, foot)	x 1^	
ohm (mil, foot)	microhm-centimeters	x 0.166 24	*x 0.17*
	microhm-inches	x 0.065 450	*x 0.065*
	ohm-meters	x 1.662 4 E-09	*x 1.7 E-09*
ohm, absolute	ohms, international	x 0.999 51	*x 1*
ohm-centimeter	ohm-meters	x 0.01^	
ohm-circular mil per foot	ohm-meters	x 1.662 4 E-09	*x 1.7 E-09*
	ohm-square millimeters per meter	x 0.001 662 4	*x 0.001 7*
ohm-meter	microhm-centimeters	x 1^E+08	
	microhm-inches	x 500 E+07/127^ or x 3.937 0 E+07	*x 4 E+07*
	ohms (mil, foot)	x 6.015 3 E+08	*x 6 E+08*
	ohm-centimeters	x 100^	
	ohm-circular mils per foot	x 6.015 3 E+08	*x 6 E+08*
ohm-square millimeter per meter	ohm-circular mils per foot	x 601.53	*x 600*
permittivity	farad per meter	x 1^	
quantum of charge	coulombs	x 1.602 1 E-19	*x 1.6 E-19*
ratio of two current levels with equal resistances	decibels	x 20 log of ratio	
ratio of two power levels	nepers	x 0.5 ln of ratio	
ratio of two voltage levels with equal resistances	decibels	x 20 log of ratio	
resistance, specific	resistivity	x 1^	
siemens	mho	x 1^	
	abmhos	x 1^E-09	
	statmhos	x 8.987 6 E+11	*x 9 E+11*
specific charge	ratio of electrical particle charge to its mass	x 1^	
statampere	amperes	x 3.335 6 E-10	*x 3.3 E-10*

TABLE 3
Measurement Conversions, By Group

All measurement units are US, unless otherwise noted. All number denominations "billion" and higher are US, unless otherwise noted.

CONVERT	TO EQUIVALENT	BY PRECISELY	OR WITHIN ± 5.0 %
statcoulomb	abcoulombs	x 3.335 6 E-11	*x 3.3 E-11*
	ampere-hours	x 9.265 7 E-14	*x 9 E-14*
	coulombs	x 3.335 6 E-10	*x 3.3 E-10*
	faradays (based on carbon-12)	x 3.457 1 E-15	*x 3.5 E-15*
statfarad	farads	x 1.112 7 E-12	*x 1.1 E-12*
stathenry	henrys	x 8.987 6 E+11	*x 9 E+11*
statmho	siemens	x 1.112 7 E-12	*x 1.1 E-12*
statohm	ohms	x 8.987 6 E+11	*x 9 E+11*
statvolt	volts	x 299.79	*x 300*
transfer ratio (for transformers)	ratio of number of turns in secondary winding to turns in primary winding	x 1^	
var	volt-ampere reactive	x 1^	
volt	abvolts	x 1^E+08	
	EMU of electric potential	x 1^E+08	
	ESU of electric potential	x 0.003 335 6	*x 0.003 3*
	statvolts	x 0.003 335 6	*x 0.003 3*
volt per centimeter	volts per inch	x 2.54^	*x 2.5*
volt per inch	volts per centimeter	x 50/127^ or x 0.393 70	*x 0.4*
volt, absolute	volts, international	x 0.999 67	*x 1*
watt, absolute	watts, international	x 0.999 84	*x 1*
ENERGY CONSUMPTION			
Btu (IT) per kilowatt-hour (power-plant heat rate)	joules per kilowatt-hour	x 1 055.1	*x 1 100*
joule per kilowatt-hour (power-plant heat rate)	Btu (IT) per kilowatt-hour	x 9.478 2 E-04	*x 9.5 E-04*
kilojoule per kilowatt-hour (power-plant heat rate)	Btu (IT) per kilowatt-hour	x 0.947 82	*x 0.95*
kilometer per liter (vehicle fuel economy)	miles per gallon	x 2.352 1	*x 2.4*
mile per gallon (vehicle fuel economy)	kilometers per liter	x 0.425 14	*x 0.43*
mpg	miles per gallon	x 1^	
ENERGY; HEAT; and WORK (also see THERMAL CONDUCTANCE; THERMAL CONDUCTIVITY; and THERMAL POWER PER UNIT AREA)			
British thermal unit	Btu	x 1^	
Btu	British thermal unit	x 1^	
	Centigrade heat units	x 5/9 or x 0.555 56	*x 0.56*
Btu (IT)	calories (IT)	x 252.00	*x 250*
	ergs	x 1.055 1 E+10	*x 1.1 E+10*
	foot-pounds (force)	x 778.17	*x 800*
	horsepower-hours	x 3.930 1 E-04	*x 4 E-04*
	joules	x 1 055.1	*x 1 100*
	kilocalories (IT)	x 0.252 00	*x 0.25*
	kilowatt-hours	x 2.930 7 E-04	*x 3 E-04*
	liter-atmospheres, standard	x 10.413	*x 10*
	meter-kilograms (force)	x 107.59	*x 110*
Btu (thermochemical)	joules	x 1 054.4	*x 1 100*
Btu (@ 39 deg. F)	joules	x 1 059.7	*x 1 100*

TABLE 3
Measurement Conversions, By Group

All measurement units are US, unless otherwise noted. All number denominations "billion" and higher are US, unless otherwise noted.

CONVERT	TO EQUIVALENT	BY PRECISELY	OR WITHIN ± 5.0 %
Btu (@ 59 deg. F)	joules	x 1 054.8	*x 1 100*
Btu (@ 60 deg. F)	joules	x 1 054.7	*x 1 100*
Btu (@ 60.5 deg. F, Can.) [4]	joules	x 1 054.6	*x 1 100*
Btu, mean (for range of 0 to 100 deg.C)	joules	x 1 055.9	*x 1 100*
calorie	kilocalories	/1 000^ or x 0.001^	
Calorie (capitalized)	kilocalories (usually, also see "calorie" (not capitalized))	x 1	
calorie [5]	See "kilocalorie"		
calorie (IT)	Btu (IT)	x 0.003 968 3	*x 0.004*
	calories, thermochemical	x 1.000 7	*x 1*
	foot-pounds (force)	x 3.088 0	*x 3*
	joules	x 4.186 8^	*x 4*
	meter-kilograms (force)	x 0.426 93	*x 0.43*
calorie (not capitalized)	calories (usually, also see "Calorie (capitalized)")	x 1^	
calorie (@ 15 deg. C)	joules	x 4.185 8	*x 4*
calorie (@ 20 deg. C)	joules	x 4.181 9	*x 4*
calorie, gram	calorie	x 1^	
calorie, kilogram	kilocalorie	x 1^	
calorie, large	kilocalorie	x 1^	
calorie, mean (for range 0 to 100 deg. C)	joules	x 4.190 0	*x 4*
calorie, Ostwald	kilocalorie	/10^ or x 0.1^	
calorie, small	calorie, gram	x 1^	
calorie, thermochemical	calories (IT)	x 0.999 33	*x 1*
	joules	x 4.184^	*x 4*
calory	calorie	x 1^	
centigrade heat unit	Btu	x 1.8	
	calories (IT)	x 453.59	*x 450*
	Chu	x 1^	
	joules	x 1 899.1	*x 1 900*
	meter-kilograms (force)	x 193.65	*x 200*
Chu	centigrade heat unit	x 1^	
dekatherm	Btu	x 1^E+06	
dyne-centimeter	foot-pounds (force)	x 7.375 6 E-08	*x 7.5 E-08*
	meter-kilograms (force)	x 1.019 7 E-08	*x 1 E-08*
electron-volt	atomic mass units (equivalent mass)	x 6.022 5 E+23	*x 6 E+23*
	ergs	x 1.602 2 E-12	*x 1.6 E-12*
energy density	joule per cubic meter	x 1^	
entropy	joule per kelvin	x 1^	
entropy, specific	ratio of entropy of a substance to its mass	x 1^	
erg	electron-volts	x 6.241 5 E+11	*x 6 E+11*
	grams (equivalent mass)	x 1.112 6 E-21	*x 1.1 E-21*
	joules	x 1^E-07	
food	See "gram of" and "ounce of"		
foot-pound (force)	Btu (IT)	x 0.001 285 1	*x 0.001 3*
	calories (IT)	x 0.323 83	*x 0.32*
	horsepower-hours	x 500 E-07/99^ or x 5.050 5 E-07	*x 5 E-07*

4 The Btu (@ 60.5 deg. F) is used by the Canadian petroleum and natural-gas industry.

5 The energy value of food and drink is customarily stated in "calories", but the technically correct measuring unit is "kilocalories", which is sometimes called "large calories". (1 kilocalorie = 1 000 calories) A person on a reducing diet may take 1 100 "calories" per day, but, in truth, he is taking 1 100 kilocalories, or 1.1 million (1 100 000) calories, or 1.1 megacalories.

TABLE 3
Measurement Conversions, By Group

All measurement units are US, unless otherwise noted. All number denominations "billion" and higher are US, unless otherwise noted.

CONVERT	TO EQUIVALENT	BY PRECISELY	OR WITHIN ± 5.0 %
	horsepower-hours, metric	x 5.120 6 E-07	*x 5 E-07*
	joules	x 1.355 8	*x 1.4*
	kilocalories	x 3.238 3 E-04	*x 3.2 E-04*
	kilowatt-hours	x 3.766 2 E-07	*x 3.8 E-07*
	liter-atmospheres	x 0.013 381	*x 0.013*
	meter-kilograms (force)	x 0.138 25	*x 0.14*
foot-poundal	joules	x 0.042 140	*x 0.042*
gram of carbohydrate	kilocalories [5]	x 4	
gram of fat	kilocalories [5]	x 9	
gram of protein	kilocalories [5]	x 4	
gram (force)-centimeter	dyne-centimeters	x 980.67	*x 1 000*
	foot-pounds (force)	x 7.233 0 E-05	*x 7 E-05*
heat capacity	joule per kelvin	x 1^	
horsepower-hour	foot-pounds (force)	x 1.98^ E+06	*x 2 E+06*
	horsepower-hours, metric	x 1.013 9	*x 1*
	joules	x 2.684 5 E+06	*x 2.7 E+06*
	kilocalories (IT)	x 641.19	*x 640*
	kilowatt-hours	x 0.745 70_	*x 0.75*
	liter-atmospheres, standard	x 26 494	*x 26 000*
	megajoules	x 2.684 5	*x 2.7*
	meter-kilograms (force)	x 2.737 4 E+05	*x 2.7 E+05*
horsepower-hour, metric	foot-pounds (force)	x 1.952 9 E+06	*x 2 E+06*
	horsepower-hours	x 0.986 32	*x 1*
	joules	x 2.647 8 E+06	*x 2.6 E+06*
	kilocalories (IT)	x 632.42	*x 630*
	kilowatt-hours	x 0.735 50	*x 0.74*
	liter-atmospheres, standard	x 26 132	*x 26 000*
	meter-kilograms (force)	x 2.700 0 E+05	*x 2.7 E+05*
joule	Btu (IT)	x 9.478 2 E-04	*x 9.5 E-04*
	Btu (@ 39 deg. F)	x 9.436 9 E-04	*x 9.4 E-04*
	Btu (@ 59 deg. F)	x 9.480 5 E-04	*x 9.5 E-04*
	Btu (@ 60 deg. F)	x 9.481 5 E-04	*x 9.5 E-04*
	Btu, mean	x 9.470 9 E-04	*x 9.5 E-04*
	Btu, thermochemical	x 9.484 5 E-04	*x 9.5 E-04*
	calories (IT)	x 0.238 85	*x 0.24*
	calories (@ 15 deg. C)	x 0.238 90	*x 0.24*
	calories (@ 20 deg. C)	x 0.239 13	*x 0.24*
	calories, mean	x 0.238 66	*x 0.24*
	calories, thermochemical	x 0.239 01	*x 0.24*
	centigrade heat units	x 5.265 6 E-04	*x 5.3 E-04*
	electronvolts	x 6.241 5 E+18	*x 6 E+18*
	foot-pounds (force)	x 0.737 56	*x 0.74*
	grams (equivalent mass)	x 1.112 7 E-14	*x 10/9 E-14*
	horsepower-hours	x 3.725 1 E-07	*x 3.7 E-07*
	horsepower-hours, metric	x 3.776 7 E-07	*x 3.8 E-07*
	kilocalories (IT)	x 2.388 5 E-04	*x 2.4 E-04*
	kilocalories, mean	x 2.386 6 E-04	*x 2.4 E-04*
	kilocalories, thermochemical	x 2.390 1 E-04	*x 2.4 E-04*

[5] The energy value of food and drink is customarily stated in "calories", but the technically correct measuring unit is "kilocalories", which is sometimes called "large calories". (1 kilocalorie = 1 000 calories) A person on a reducing diet may take 1 100 "calories" per day, but, in truth, he is taking 1 100 kilocalories, or 1.1 million (1 100 000) calories, or 1.1 megacalories.

TABLE 3
Measurement Conversions, By Group

All measurement units are US, unless otherwise noted. All number denominations "billion" and higher are US, unless otherwise noted.

CONVERT	TO EQUIVALENT	BY PRECISELY	OR WITHIN ± 5.0 %
	liter-atmospheres, standard	x 0.009 869 2	*x 0.01*
	meter-kilograms (force)	x 0.101 97	*x 0.1*
	meter-newton	x 1^	
	quads (energy)	x 9.479 E-19	*x 9.5 E-19*
	therms (EEC)	x 9.478 1 E-09	*x 9.5 E-09*
	therms (US)	x 9.480 4 E-09	*x 9.5 E-09*
	watt-second	x 1^	
kilocalorie	Btu (IT)	x 3.968 3	*x 4*
	calories	x 1 000^	
kilocalorie (IT)	Btu (IT)	x 3.968 3	*x 4*
	foot-pounds (force)	x 3 088.0	*x 3 000*
	horsepower-hours	x 0.001 559 6	*x 0.001 6*
	horsepower-hours, metric	x 0.001 581 2	*x 0.001 6*
	joules	x 4 186.8^	*x 4 000*
	kilowatt-hours	x 0.001 163^	*x 0.001 2*
	liter-atmospheres, standard	x 41.321	*x 40*
	meter-kilograms (force)	x 426.93	*x 430*
kilocalorie (IT) (for foods and biological heat output) [5]	joules	x 4 186.8^	*x 4 000*
kilocalorie (thermochemical)	joules	x 4 184^	*x 4 000*
kilocalorie, mean (over range 0 to 100 deg. C)	joules	x 4 190.0	*x 4 000*
kilogram-calorie	kilocalorie	x 1^	
kilowatt-hour	Btu (IT)	x 3 412.1	*x 3 400*
	calories (IT)	x 8.598 5 E+05	*x 9 E+05*
	ergs	x 3.6^E+13	
	foot-pounds (force)	x 2.655 2 E+06	*x 2.7 E+06*
	horsepower-hours	x 1.341 0	*x 1.3*
	horsepower-hours, metric	x 1.359 6	*x 1.4*
	joules	x 3.6^E+06	
	kilocalories (IT)	x 859.85	*x 900*
	liter-atmospheres, standard	x 35 529	*x 36 000*
	megajoules	x 3.6^	
	meter-kilograms (force)	x 3.671 0 E+05	*x 3.7 E+05*
liter-atmosphere, standard	foot-pounds (force)	x 74.733	*x 75*
	horsepower-hours	x 3.774 4 E-05	*x 3.8 E-05*
	horsepower-hours, metric	x 3.826 8 E-05	*x 3.8 E-05*
	joules	x 101.32	*x 100*
	kilocalories (IT)	x 0.024 201	*x 0.024*
	kilograms (force)	x 10.332	*x 10*
	kilowatt-hours	x 2.814 6 E-05	*x 2.8 E-05*
megajoule	horsepower-hours	x 0.373 13	*x 0.37*
	kilowatt-hours	x 5/18^ or x 0.277 78	*x 0.28*
Mercalli scale (for earthquakes)	See APPENDIX, "earthquakes"		
meter-kilogram (force)	Btu (IT)	x 0.009 294 9	*x 0.009*
	calories (IT)	x 2.342 2	*x 2.3*

[5] The energy value of food and drink is customarily stated in "calories", but the technically correct measuring unit is "kilocalories", which is sometimes called "large calories". (1 kilocalorie = 1 000 calories) A person on a reducing diet may take 1 100 "calories" per day, but, in truth, he is taking 1 100 kilocalories, or 1.1 million (1 100 000) calories, or 1.1 megacalories.

TABLE 3
Measurement Conversions, By Group

All measurement units are US, unless otherwise noted. All number denominations "billion" and higher are US, unless otherwise noted.

CONVERT	TO EQUIVALENT	BY PRECISELY	OR WITHIN ± 5.0 %
	Centigrade heat units	x 0.005 163 8	*x 0.005*
	foot-pounds (force)	x 7.233 0	*x 7*
	horsepower-hours	x 3.653 0 E-06	*x 3.7 E-06*
	horsepower-hours, metric	x 3.703 7 E-06	*x 3.7 E-06*
	joules	x 9.806 6	*x 10*
	kilocalories (IT)	x 0.002 342 3	*x 0.002 3*
	kilowatt-hours	x 2.724 1 E-06	*x 2.7 E-06*
	liter-atmospheres, standard	x 0.096 784	*x 0.1*
meter-newton	joule	x 1^	
metric horsepower-hour	See "horsepower-hour, metric"		
mev	million electron-volts	x1^	
molar energy	joule per mole	x 1^	
molar entropy	joule per mole-kelvin	x 1^	
molar heat capacity	joule per mole-kelvin	x 1^	
ounce of carbohydrate	kilocalories [5]	x 110 (approx.)	
ounce of fat	kilocalories [5]	x 260 (approx.)	
ounce of liquor, 80 proof	kilocalories [5]	x 65	
ounce of protein	kilocalories [5]	x 110 (approx.)	
photon	quantum of electromagnetic energy	x 1^	
quad (energy)	Btu (IT)	x 1^E+15	
	joules	x 1.055 E+18	
quantum of charge	See ELECTRICITY		
Richter magnitude scale (for earthquakes)	Also see LENGTH; TORQUE; and APPENDIX, "earthquakes"		
Richter scale (for earthquakes) increase of 0.1 magnitude	freed-energy increase	x 10 to power 0.15 or x 1.4	
Richter scale (for earthquakes) increase of one magnitude	freed-energy increase	x 10 to power 1.5 or x 32	
solar constant (for radiation falling perpendicularly to the earth's surface)	calories per minute-square centimeter	x 1.94 (mean)	*x 2*
specific energy	joule per kilogram	x 1^	
specific entropy	joule per kilogram-kelvin	x 1^	
specific heat capacity	joule per kilogram-kelvin	x 1^	
therm (AGA)	Btu (IT)	x 1.000 00 E+05	*x 100 000*
therm (EEC)	Btu (IT)	x 1.000 00 E+05	*x 100 000*
	calories	x 2.520 0 E+07	*x 2.5 E+07*
	joules	x 1.055 1 E+08	*x 1.1 E+08*
therm (US)	Btu (IT)	x 9.997 61 E+04	*x 100 000*
	calories	x 2.519 8 E+07	*x 2.5 E+07*
	joules	x 1.054 8 E+08	*x 1.1 E+08*
therme	therm	x 1^	
ton (energy)	Also see FORCE; MASS; and VOLUME		
ton (explosive energy of one ton of TNT)	joules	x 4.184 E+09 [6]	*x 4 E+09*

[5] The energy value of food and drink is customarily stated in "calories", but the technically correct measuring unit is "kilocalories", which is sometimes called "large calories". (1 kilocalorie = 1 000 calories) A person on a reducing diet may take 1 100 "calories" per day, but, in truth, he is taking 1 100 kilocalories, or 1.1 million (1 100 000) calories, or 1.1 megacalories.

[6] The ton of TNT energy is defined, not measured.

TABLE 3
Measurement Conversions, By Group

All measurement units are US, unless otherwise noted. All number denominations "billion" and higher are US, unless otherwise noted.

CONVERT	TO EQUIVALENT	BY PRECISELY	OR WITHIN ± 5.0 %
	See FORCE, "megaton"		
watt-hour	Btu (IT)	x 3.412 1	*x 3.4*
	foot-pounds (force)	x 2 655.2	*x 2 700*
	horsepower-hours	x 0.001 341 0	*x 0.001 3*
	joules	x 3 600^	
	kilocalories	x 0.859 85	*x 0.9*
	meter-kilograms (force)	x 367.10	*x 370*
watt-second	joules	x 1^	
Wobbe index (for flammable gas)	ratio of heat of combustion to specific gravity	x 1^	
ENERGY PER UNIT AREA			
langley (solar radiation)	calorie per square centimeter	x 1^	
	joules per square meter	x 41 840^	*x 40 000*
FLOW RATE (of material)			
acre-foot per day	acre-inches per hour	/2^ or x 0.5^	
	gallons per minute	x 226.29	*x 220*
acre-foot per hour	cubic feet per hour	x 43 560^	*x 44 000*
	gallons per minute	x 5 430.9	*x 5 400*
barrel (42 gallons) per day	barrels (42 gallons) per hour	/24^ or x 0.041 667	*x 0.04*
	cubic feet per hour	x 0.233 94	*x 0.23*
	cubic feet per minute	x 0.003 899 0	*x 0.004*
	cubic feet per second	x 6.498 4 E-05	*x 6.5 E-05*
	gallons per hour	x 1.750 0	*x 1.8*
	gallons per minute	x 7/240^ or x 0.029 167	*x 0.03*
	liters per second	x 0.001 840 1	*x 0.001 8*
barrel (42 gallons) per hour	barrels (42 gallons) per day	x 24^	
	cubic feet per hour	x 5.614 6	*x 5.6*
	cubic feet per minute	x 0.093 576	*x 0.09*
	cubic feet per second	x 0.001 559 6	*x 0.001 5*
	gallons per hour	x 42^	
	gallons per minute	x 0.7^	
	liters per second	x 0.044 163	*x 0.044*
bpd	barrel per day	x 1^	
bph	barrel per hour	x 1^	
bpm	barrel per minute	x 1^	
cfd	cubic foot per day	x 1^	
cfh	cubic foot per hour	x 1^	
cfm	cubic foot per minute	x 1^	
cfs	cubic foot per second	x 1 ^	
cubic foot per hour	acre-feet per hour	x 2.295 7 E-05	*x 2.3 E-05*
	barrels (42 gallons) per day	x 4.274 6	*x 4.3*
	barrels (42 gallons) per hour	x 0.178 11	*x 0.18*
	cubic feet per minute	/60^ or x 0.016 667	*x 0.017*
	cubic feet per second	/3 600^ or x 2.777 8 E-04	
	gallons per hour	x 7.480 5	*x 7.5*
	gallons per minute	x 0.124 68	*x 0.12*
	liters per second	x 0.007 865 8	*x 0.008*
cubic foot per minute	barrels (42 gallons) per day	x 256.47	*x 260*
	barrels (42 gallons) per hour	x 10.686	*x 11*
	cubic feet per hour	x 60^	

TABLE 3
Measurement Conversions, By Group

All measurement units are US, unless otherwise noted. All number denominations "billion" and higher are US, unless otherwise noted.

CONVERT	TO EQUIVALENT	BY PRECISELY	OR WITHIN ± 5.0 %
	cubic feet per second	/60^ or x 0.016 667	x 0.017
	cubic meters per second	x 4.719 5 E-04	x 4.7 E-04
	gallons per hour	x 448.83	x 450
	gallons per minute	x 7.480 5	x 7.5
	liters per second	x 0.471 95	x 0.47
cubic foot per second	barrels (42 gallons) per day	x 15 388	x 15 000
	barrels (42 gallons) per hour	x 641.19	x 640
	cubic feet per hour	x 3 600^	
	cubic feet per minute	x 60^	
	cubic meters per second	x 0.028 317	x 0.028
	cubic yards per minute	x 20/9^ or x 2.222 2	x 2.2
	gallons per hour	x 26 930	x 27 000
	gallons per minute	x 448.83	x 450 or x 4 000/9
	liters per second	x 28.317	x 28
	million gallons per day	x 0.646 32	x 0.65
cubic inch per minute	cubic meters per second	x 2.731 2 E-07	x 2.7 E-07
cubic meter per hour	gallons per minute	x 4.402 9	x 40/9
cubic meter per second	cubic feet per minute	x 2 118.9	x 2 100
	cubic feet per second	x 35.315	x 35
	cubic inches per minute	x 3.661 4 E+06	x 3.7 E+06
	cubic yards per minute	x 78.477	x 80
	gallons per day	x 2.282 4 E+07	x 2.3 E+07
	gallons per hour	x 9.510 2 E+05	x 9.5 E+05
	gallons per minute	x 15 850	x 16 000
cubic yard per minute	cubic feet per second	x 9/20^ or x 0.45^	x 4/9
	cubic meters per second	x 0.012 743	/80 or x 0.013
	gallons per second	x 3.366 2	x 10/3 or x 3.4
	liters per second	x 12.743	x 25/2 or x 13
C_v (valve flow coefficient)	number of gallons per minute [7]	x 1^	
gallon per day	cubic meters per second	x 4.381 3 E-08	x 4.4 E-08 or x 4/9 E-08
gallon per hour	barrels (42 gallons) per day	x 0.571 43	x 0.57
	barrels (42 gallons) per hour	x 0.023 810	x 0.024
	cubic feet per hour	x 0.133 68	x 0.13
	cubic feet per minute	x 0.002 228 0	x 0.002 2
	cubic feet per second	x 3.713 3 E-05	x 3.7 E-05
	cubic meters per second	x 1.051 5 E-06	x 1.1 E-06
	gallons per minute	/60^ or x 0.016 667	x 0.017
	liters per second	x 0.001 051 5	x 0.001 1
gallon per minute	acre-feet per hour	x 1.841 3 E-04	x 1.8 E-04
	barrels (42 gallons) per day	x 34.286	x 34
	barrels (42 gallons) per hour	x 1.428 6	x 1.4
	cubic feet per hour	x 8.020 8	x 8
	cubic feet per minute	x 0.133 68	x 0.13
	cubic feet per second	x 0.002 228 0	x 0.002 2
	cubic meters per second	x 6.309 0 E-05	x 6.3 E-05
	gallons per hour	x 60^	
	liters per minute	x 3.785 4	x 3.8
	liters per second	x 0.063 090	x 0.063

[7] C_v is the number of US gallons per minute of water at 60 deg. F flowing through a valve when the pressure drop across the valve is one pound per square inch under stated conditions of pressure and valve opening.

TABLE 3
Measurement Conversions, By Group

All measurement units are US, unless otherwise noted. All number denominations "billion" and higher are US, unless otherwise noted.

CONVERT	TO EQUIVALENT	BY PRECISELY	OR WITHIN ± 5.0 %
	tons of water per day	x 6.008 6 (@ 4 deg Celsius)	*x 6*
gpd	gallon per day	x 1^	
gph	gallon per hour	x 1^	
gpm	gallon per minute	x 1^	
gps	gallon per second	x 1^	
inch, miner's	cubic feet per minute	x 1.5 (usually)	
kilogram per second	pounds per hour	x 7 936.6	*x 8 000*
	pounds per minute	x 132.28	*x 130*
	pounds per second	x 2.204 6	*x 2.2*
	tons, short per hour	x 3.968 3	*x 4*
liter per minute	cubic feet per second	x 5.885 8 E-04	*x 6 E-04*
	gallons per minute	x 0.264 17	*x 0.26*
	gallons per second	x 0.004 402 9	*x 0.004 4*
	gallons, Imperial per second	x 0.003 666 2	*x 0.003 7*
liter per second	barrels (42 gallons) per day	x 543.44	*x 540*
	barrels (42 gallons) per hour	x 22.643	*x 23*
	cubic feet per hour	x 127.13	*x 130*
	cubic feet per minute	x 2.118 9	*x 2.1*
	cubic feet per second	x 0.035 315	*x 0.035*
	gallons per hour	x 951.02	*x 950*
	gallons per minute	x 15.850	*x 16*
million gallons per day	cubic feet per second	x 1.547 2	*x 1.5*
miner's inch	See "inch, miner's"		
pound of water per minute (@ 4 deg. C)	cubic feet per second	x 2.669 8 E-04	*x 2.7 E-04*
pound per hour	kilograms per second	x 1.259 9 E-04	*x 1.3 E-04*
pound per minute	kilograms per second	x 0.007 559 9	*x 0.007 6*
pound per second	kilograms per second	x 0.453 59	*x 0.45*
ton of material per day	pounds of material per hour	x 250/3^ or x 83.333	*x 80*
ton of water per day (@ 4 deg. C)	cubic feet per hour	x 1.334 9	*x 1.3*
	gallons per minute	x 0.166 43	*x 0.17*
ton, short per hour	kilograms per second	x 0.252 00	*x 0.25*
FLUIDITY (see VISCOSITY)			
FORCE (also see MASS)			
Beaufort scale (for wind)	See APPENDIX, "wind speeds"		
dyne	gram-centimeter per second per second	x 1^	
	joule per centimeter	x 1^E-07	
	joules per meter	x 1^E-05	
	newtons	x 1^E-05	
	pounds (force)	x 2.248 1 E-06	*x 2.2 E-06*
force of gravity	weight	x 1^	
gram (force)	dynes	x 980.67	*x 1 000*
	newtons	x 9 806.7	*x 10 000*
	poundals	x 0.070 932	*x 0.07*
kilogram (force)	newtons	x 9.806 7	*x 10*
kilogram (force) per meter	pounds (force) per foot	x 0.671 97	*x 0.7*
kilonewton	tons, short (force)	x 0.112 40	*x 0.11*

TABLE 3
Measurement Conversions, By Group

All measurement units are US, unless otherwise noted. All number denominations "billion" and higher are US, unless otherwise noted.

CONVERT	TO EQUIVALENT	BY PRECISELY	OR WITHIN ± 5.0 %
kilopond	kilogram (force, @ standard gravity) [8]	x 1^	
kip (kilopound, force)	newtons	x 4 448.2	*x 4 400*
	pounds (force)	x 1 000^	
megaton (explosive force)	tons of TNT	x 1^E+06	
megaton (explosive force)	Also see ENERGY, "ton (explosive ...)"		
newton	dynes	x 1^E+05	
	kilograms (force)	x 0.101 97	*x 0.1*
	kiloponds	x 0.101 97	*x 0.1*
	kips	x 2.248 1 E-04	*x 2.2 E-04*
	ounces (force)	x 3.596 9	*x 3.6*
	poundals	x 7.233 0	*x 7*
	pounds (force)	x 0.224 81	*x 0.22*
	tons, short (force)	x 1.124 0 E-04	*x 1.1 E-04*
ounce (force)	newtons	x 0.278 01	*x 0.28*
pound (force)	dynes	x 4.448 2 E+05	*x 4.4 E+05*
	grams (force)	x 453.59	*x 450*
	kips	/1 000^ or x 0.001^	
	newtons	x 4.448 2	*x 4.4*
	poundals	x 32.174	*x 32*
poundal	grams (force)	x 14.098	*x 14*
	newtons	x 0.138 26	*x 0.14*
	pound (mass)-foot per second per second	x 1^	
	pounds (force)	x 0.031 081	*x 0.03*
specific impulse	ratio of rocket thrust to propellant consumption rate	x 1^	
specific thrust	specific impulse	x 1^	
ton (force)	Also see ENERGY; MASS; and VOLUME		
ton, short (force)	kilonewtons	x 8.896 4	*x 9*
	newtons	x 8 896.4	*x 9 000*
weight (force)	mass	x g (see ACCELERATION, "g")	
FORCE PER UNIT AREA (see PRESSURE)			
FORCE PER UNIT LENGTH and SPRING RATE			
kilogram (force) per meter	pounds (force) per foot	x 0.671 97	*x 0.7*
newton per meter	pounds (force) per foot	x 0.068 522	*x 0.07*
	pounds (force) per inch	x 0.571 01	*x 0.57*
pound (force) per foot	kilograms (force) per meter	x 1.488 2	*x 1.5*
	newtons per meter	x 14.594	*x 15*
pound (force) per inch	newtons per meter	x 175.13	*x 180*
FREQUENCY (see TIME)			
GRAVITY, SPECIFIC (see SPECIFIC GRAVITY)			
HEAT (see ENERGY)			

[8] Standard conditions for the acceleration of gravity are: 9.806 65 meters per second per second = 32.174 0 feet per second per second, at sea level and latitude 45 degrees. At other locations, the acceleration may differ within a span of more than 0.5 percent of the standard value.

TABLE 3
Measurement Conversions, By Group

All measurement units are US, unless otherwise noted. All number denominations "billion" and higher are US, unless otherwise noted.

CONVERT	TO EQUIVALENT	BY PRECISELY	OR WITHIN ± 5.0 %
LENGTH and LINEAR DISTANCE *The "mile" is the "statute mile". The statute mile and the nautical mile are US, unless otherwise noted.*			
agate line (in newspaper advertisements)	one column wide, 1/14-inch deep	x 1	
agate (printer's type)	points (printer's)	x 5.5	
angstrom	centimeters	x 1^E-08	
	inches	x 3.937 0 E-09	*x 4 E-09*
	meters	x 1^E-10	
	microns	x 1^E-04	
arpent (French land length, Queb.)	feet (French land measure, Queb.)	x 180^	
astronomical unit	kilometers	x 1.496 0 E+08	*x 1.5 E+08*
	light-year	x 1.581 3 E-05	*x 1.6 E-05*
	meters	x 1.496 0 E+11	*x 1.5 E+11*
	parsec	x 4.848 1 E-06	*x 5 E-06*
bolt (for cloth) (US)	meters	x 18.3 to x 36.6 (usually)	
	yards	x 20 to 40 (usually)	
cable	See "cable length"		
cable length	fathoms	x 120^	
	feet	x 720^	
	meters	x 219.46	*x 220*
	miles	x 3/22^ or x 0.136 36	*x 0.14*
cable length (Brit.)	feet	x 608^	*x 600*
	miles	x 0.115 15	*x 0.12*
centimeter	angstroms	x 1^E+08	
	feet	x 0.032 808	*x 0.032*
	inches	/2.54^ or x 0.393 70	*x 0.4*
	meters	/100^ or x 0.01^	
	miles	x 6.213 7 E-06	*x 6 E-06*
	mils	x 393.70	*x 400*
	yards	x 0.010 936	*x 0.011*
chain (Can.)	chain, surveyor's (US)	x 1^	
chain (for football, US)	yards	x 10^	
chain, engineer's	feet	x 100^	
	links	x 100^	
chain, Gunter's	chain, surveyor's	x 1^	
chain, Ramden's	chain, engineer's	x 1^	
chain, surveyor's	feet	x 66^	
	links	x 100^	
	meters	x 20.117	*x 20*
	miles	/80^ or x 0.012 5^	*x 0.013*
	rods (length)	x 4^	
cubit (ancient)	inches	x 17 to x 22 (approx.)	
decimeter	feet	x 0.328 08	*x 0.33*
	inches	x 3.937 0	*x 4*
deep six	fathoms	x 6^	
dekameter	rods (length)	x 1.988 4	*x 2*
ell (Eng.)	centimeters	x 114.3^	*x 110*
	inches	x 45^	
em, pica	centimeters	x 0.421 75	*x 0.42*
	inches	x 0.166 04	*x 0.17*

TABLE 3
Measurement Conversions, By Group

All measurement units are US, unless otherwise noted. All number denominations "billion" and higher are US, unless otherwise noted.

CONVERT	TO EQUIVALENT	BY PRECISELY	OR WITHIN ± 5.0 %
fathom	feet	x 6^	
	meters	x 1.828 8	*x 1.8*
fermi	meters	x 1^E-15	
foot	centimeters	x 30.48^	*x 30*
	chains, engineer's	/100^ or x 0.01^	
	chains, surveyor's	/66^ or x 0.015 152	*x 0.015*
	fathoms	/6^ or x 0.166 67	*x 0.17*
	foot, statute	x 1^	
	foot, survey	x 1	
	furlongs	/660^ or x 0.001 515 2	*x 0.001 5*
	inches	x 12^	
	links, engineer's	x 1^	
	links, surveyor's	/0.66^ or x 1.515 2	*x 1.5*
	meters	x 0.304 8^	*x 0.3*
	miles	/5 280^ or x 1.893 9 E-04	*x 1.9 E-04*
	miles, nautical (intl.)	x 1.645 8 E-04	*x 1.6 E-04*
	mils (length)	x 12 000^	
	perch (length)	/16.5^ or x 0.060 606	*x 0.06*
	rods (length)	/16.5^ or x 0.060 606	*x 0.06*
	yards	/3^ or x 0.333 33	*x 0.33*
foot (Brit.)	foot (US)	x 1^	
foot (Can.)	foot (US)	x 1^	
foot (intl.)	foot, statute (US)	x 1^	
	foot, survey (US)	x 1	
	meters	x 0.304 8^	*x 0.3*
foot, Paris (French land measure, Queb.)	foot (French measure, Queb.)	x 1^	
	inches (English measure, Queb.)	x 12.789^	*x 13*
foot, statute	meters	x 0.304 8^	*x 0.3*
foot, survey (for land measure) [9]	meters	x 1 200/3 937^ or x 0.304 80	*x 0.3*
furlong	feet	x 660^	
	meters	x 201.17	*x 200*
	miles	/8^ or x 0.125^	*x 0.13*
	rods	x 40^	
furlong (Can.)	furlong (US)	x 1^	
hand (for height of horses)	centimeters	x 10.16^	*x 10*
	inches	x 4^	
inch	angstroms	x 2.54^E+08	*x 2.5 E+08*
	centimeters	x 2.54^	*x 2.5*
	feet	/12^ or x 0.083 333	*x 0.08*
	hands (for height of horses)	/4 or x 0.25^	
	meters	x 0.025 4^	*x 0.025*
	microns	x 2.54^E+04	*x 2.5 E+04*
	miles	/63 360^ or x 1.578 3 E-05	*x 1.6 E-05*
	millimeters	x 25.4^	*x 25*
	mils (length)	x 1 000^	
	points (printer's)	x 72.281	*x 72*
	yards	/36^ or x 0.027 778	*x 0.028*
inch (Brit.)	inch (US)	x 1^	
inch (Can.)	inch (US)	x 1^	

[9] The length of the survey foot may be revised.

TABLE 3
Measurement Conversions, By Group

All measurement units are US, unless otherwise noted. All number denominations "billion" and higher are US, unless otherwise noted.

CONVERT	TO EQUIVALENT	BY PRECISELY	OR WITHIN ± 5.0 %
kilometer	leagues, land	x 0.207 12	x 0.2
	marathons	x 0.023 699	x 0.024
	miles	x 0.621 37	x 5/8 or x 0.6
	miles (intl.)	x 0.621 37	x 5/8 or x 0.6
	miles, nautical (intl.)	x 0.539 96	x 0.54 or x 5/9
	miles, nautical (US)	x 0.539 96	x 0.54 or x 5/9
	miles, statute (intl.)	x 0.621 37	x 5/8 or x 0.6
	miles, statute (US)	x 0.621 37	x 5/8 or x 0.6
league (Texas)	See AREA		
league, land	kilometers	x 4.828 0	x 5
	miles	x 3^	
league, marine	league, nautical	x 1^	
league, nautical	miles, nautical	x 3^	
league, sea	league, nautical	x 1^	
light-year	astronomical units	x 63 240	x 63 000
	kilometers	x 9.460 6 E+12	x 9.5 E+12
	meters	x 9.460 6 E+15	x 9.5 E+15
	miles	x 5.878 5 E+12	x 6 E+12
	parsecs	x 0.306 60	x 0.3
ligne (for buttons)	line (for buttons)	x 1^	
line (for buttons)	inches	/40^	
line (for fisherman's setline)	fathoms	x 50	
line (for whaling harpoons)	fathoms	x 150 (approx.)	
line (length, obsolete)	inches	/12^	
link (Can.)	link, surveyor's (US)	x 1^	
link, engineer's	inches	x 12^	
link, surveyor's	centimeters	x 20.117	x 20
	chains, surveyor's	/100^ or x 0.01^	
	feet	x 0.66^	
	inches	x 7.92^	x 8
	meters	x 0.201 17	x 0.2
	miles	/8 000^ or x 1.25^E-04	x 1.3 E-04
	rods (length)	/25^ or x 0.04^	
marathon	26 miles, 385 yards	x 1^	
	kilometers	x 42.195	x 42
	miles	x 26.219	x 26
meter	angstroms	x 1^E+10	
	centimeters	x 100^	
	chains, engineer's	x 0.032 808	x 0.033
	chains, surveyor's	x 0.049 710	x 0.05
	fathoms	x 0.546 81	x 0.55
	feet	x 1 250/381^ or x 3.280 8	x 3.3
	furlongs	x 0.004 971 0	x 0.005
	inches	x 39.370	x 40
	links, engineer's	x 3.280 8	x 3.3
	links, surveyor's	x 4.971 0	x 5
	microinches	x 3.937 0 E+07	x 4 E+07
	micromicrons	x 1^E+12	
	miles	x 6.213 7 E-04	x 6 E-04
	miles, nautical (intl.)	x 5.399 6 E-04	x 5.4 E-04
	miles, nautical (US)	x 5.399 6 E-04	x 5.4 E-04

TABLE 3
Measurement Conversions, By Group

All measurement units are US, unless otherwise noted. All number denominations "billion" and higher are US, unless otherwise noted.

CONVERT	TO EQUIVALENT	BY PRECISELY	OR WITHIN ± 5.0 %
	miles, statute (US)	x 6.213 7 E-04	*x 6 E-04*
	mils (length)	x 39 370	*x 40 000*
	rods (length)	x 0.198 84	*x 0.2*
	yards	x 1.093 6	*x 1.1*
metre	meter (US)	x 1^	
microinch	inches	x 1^E-06	
	meters	x 2.54^E-08	*x 2.5 E-08*
	millimeters	x 2.54^E-05	*x 2.5 E-05*
micrometer	millimeters	/1 000^ or x 0.001^	
micromicron	meters	x 1^E-12	
micron	inches	x 3.937 0 E-05	*x 4 E-05*
	meters	x 1^E-06	
mil (length)	inches	/1 000^ or x 0.001^	
	meters	x 2.54^E-05	*x 2.5 E-05*
	millimeters	x 0.025 4^	*x 0.025*
mile	centimeters	x 1.609 3 E+05	*x 1.6 E+05*
	chains, surveyor's	x 80^	
	feet	x 5 280^	*x 5 300*
	furlongs	x 8^	
	inches	x 63 360^	*x 63 000*
	kilometers	x 1.609 3	*x 1.6*
	leagues, land	/3^ or x 0.333 33	*x 0.33*
	links, engineer's	x 5 280^	*x 5 300*
	links, surveyor's	x 8 000^	
	meters	x 1 609.3	*x 1 600*
	mile, nautical	x 0.868 98	*x 13/15 or x 7/8*
	mile, statute	x 1^	
	rods (length)	x 320^	
	yards	x 1 760^	*x 1 800*
mile of line	distance between two points connected by rail line	x 1^	
mile of road	mile of line	x 1^	
mile (Brit.)	feet	x 5 000^ (obsolete)	
	feet	x 5 280^	*x 5 300*
	kilometers	x 1.609 3	*x 1.6*
	meters	x 1 609.3	*x 1 600*
	mile (US)	x 1^	
mile (Can.)	mile, statute (US)	x 1^	
mile (intl.)	feet (intl.)	x 5 280^	*x 5 300*
	kilometers	x 1.609 3	*x 1.6*
	meters	x 1 609.3	*x 1 600*
	mile (US)	x 1	
	mile, statute (US)	x 1	
mile (US)	leagues (British)	/3 or x 0.3	
mile, Admiralty (Brit.)	mile, nautical (Brit.)	x 1^	
mile, air	feet (1946 - 1954)	x 6 080.2	*x 6 100*
	feet (before 1946)	x 5 280^	*x 5 300*
	feet (since 1954)	x 6 076.1	*x 6 000*
	meters (since 1954)	x 1 852^	*x 1 900*
mile, air (US)	mile, nautical (Brit., 1946 - 1954)	x 1^	
	mile, nautical (intl., since 1954)	x 1^	
	mile, statute (US, before 1946)	x 1^	
mile, geographic	mile, nautical	x 1^	

TABLE 3
Measurement Conversions, By Group

All measurement units are US, unless otherwise noted. All number denominations "billion" and higher are US, unless otherwise noted.

CONVERT	TO EQUIVALENT	BY PRECISELY	OR WITHIN ± 5.0 %
mile, nautical	feet	x 6 076.1 (since 1959)	*x 6 000*
	feet	x 6 080.2 (obsolete)	*x 6 100*
	kilometers	x 1.852^ (since 1959)	*x 1.9*
	leagues, nautical	/3^ or x 0.333 33	*x 0.33*
	meters	x 1 852^ (since 1959)	*x 1 900*
mile, nautical (Brit.)	feet	x 6 076.1	*x 6 000*
	feet	x 6 080 (obsolete)	*x 6 000*
	kilometers	x 1.852^	*x 1.9*
	meters	x 1 853.2 (obsolete)	*x 1 900*
	meters	x 1 852^	*x 1 900*
mile, nautical (Can.)	meters	x 1 852^	
mile, nautical (intl.)	feet	x 6 076.1	*x 6 000*
	kilometers	x 1.852^	*x 1.9*
	meters	x 1 852^	*x 1 900*
	mile, nautical (US)	x 1^	
	mile, statute (US)	x 1.150 8	*x 1.2*
mile, sea	mile, nautical	x 1^	
mile, statute	feet, survey	x 5 280	*x 5 300*
	meters	x 1 609.3	*x 1 600*
	mile	x 1^	
	mile, survey	x 1	
millimeter	inches	x 0.039 370	*x 0.4*
	microinches	x 39 370	*x 40 000*
	mils (length)	x 39.370	*x 40*
myriameter (obsolete)	meters	x 10 000^	
nail (for cloth)	inches	x 2.25^	*x 2.3*
	yards	/16^ or x 0.062 5^	*x 0.06*
nanometer	microns	/1 000^ or x 0.001^	
pace, double-time	inches	x 36	
pace, geometrical	feet	x 5	
pace, military	inches	x 30	
pace, quick-time	inches	x 30	
palm	inches	x 3	
parsec	astronomical units	x 2.062 6 E+05	*x 2 E+05*
	kilometers	x 3.085 7 E+13	*x 3 E+13*
	light-years	x 3.261 6	*x 3.3*
	meters	x 3.085 7 E+16	*x 3 E+16*
	miles	x 1.917 3 E+13	*x 2 E+13*
perch	Also see AREA and VOLUME		
perch (Can.)	perch (US)	x 1^	
perch (French land length, Queb.)	feet (French land measure, Queb.)	x 18^	
perch (length)	rod (length)	x 1^	
pica (printer's)	Also see QUANTITY PER UNIT LENGTH		
	inches	x 0.166 02	*/6 or x 0.17*
	millimeters	x 4.216 9	*x 4.2*
	points (printer's)	x 12^	
point (printer's)	inches	x 0.013 835	*x 0.014*
	millimeters	x 350/996^ or x 0.351 41	*x 0.35*
point (thickness of paper)	inches	/1 000^	
pole (Can.)	pole (US)	x 1^	
pole (length)	rod (length)	x 1^	

TABLE 3
Measurement Conversions, By Group

All measurement units are US, unless otherwise noted. All number denominations "billion" and higher are US, unless otherwise noted.

CONVERT	TO EQUIVALENT	BY PRECISELY	OR WITHIN ± 5.0 %
quarter (length)	furlongs	x 2^	
	miles	/4^ or x 0.25^	
railway track gage, narrow	inches	x 36^ (usually), 30^, or 42^	
railway track gage, standard	inches	x 56.5^	*x 57*
Richter magnitude scale (for earthquakes)	Also see ENERGY; TORQUE; and APPENDIX, "earthquakes"		
Richter scale (for earthquakes) increase of 0.1 magnitude	ground-motion-amplitude increase	x 10 to power 0.1 or x 1.258 9	*x 1.3*
Richter scale (for earthquakes) increase of one magnitude	ground-motion-amplitude increase	x 10	
rod	Also see AREA and VOLUME		
	chains, surveyor's	/4^ or x 0.25^	
	feet	x 16.5^	*x 17*
	furlongs	/40^ or x 0.025^	
	links, engineer's	x 16.5^	*x 17*
	links, surveyor's	x 25^	
	meters	x 5.029 2	*x 5*
	miles	x 5/1 600^ or x 0.003 125^	*x 0.003 1*
	yards	x 5.5^	
rod (Can.)	rod (US)	x 1^	
rood (length, Brit.)	Also see AREA		
	yards	x 5.5 to x 8	
skein	feet	x 360^	
	meters	x 109.73	*x 110*
span	inches	x 9^	
standard distance (for capital ships)	yards	x 550	
standard distance (for cruisers)	yards	x 250	
swimming pool, junior olympic	meters	x 25^	
swimming pool, olympic	meters	x 50^	
twain (mark twain)	fathoms	x 2^	
unit, astronomical	See "astronomical unit"		
vara (Texas)	inches	x 33.33	*x 33*
wave number	1/meters	x 1^	
	1/wave length	x 1^	
	x 1/meters	x 1^	
yard	centimeters	x 91.44^	*x 90*
	feet	x 3^	
	inches	x 36^	
	miles	/1 760^ or x 5.681 8 E-04	*x 5.7 E-04*
yard of ale	See VOLUME		
yard of land	See AREA		
yard (before 1959)	meters	x 3 600/3 937^ or x 0.914 40	*x 0.9*
yard (British)	yard (US)	x 1^	
yard (Can.)	yard (US)	x 1^	
yard (since 1959)	meters	x 0.914 4^	*x 0.9*
year, light	See "light-year"		

TABLE 3
Measurement Conversions, By Group

All measurement units are US, unless otherwise noted. All number denominations "billion" and higher are US, unless otherwise noted.

CONVERT	TO EQUIVALENT	BY PRECISELY	OR WITHIN ± 5.0 %
LIGHT			
candela per square centimeter	lamberts	x pi or x 3.141 6	*x 22/7 or x 3*
candela per square inch	candelas per square meter	x 1 550.0	*x 1 600*
candela per square meter	candelas per square inch	x 6.451 6 E-04	*x 6.5 E-04*
	footlamberts	x 0.29 186	*x 0.3*
	stilbs	/10 000^ or x 1^E-04	
candle	candelas	x 1.02	*x 1*
candle per square centimeter	candles per square inch	x 6.451 6^	*x 6.5*
candlepower	See "spherical candlepower"		
candlepower, spherical	See "spherical candlepower"		
candle, hefner	candela	x 0.92	*x 0.9*
candle, international	candle	x 1^	
	lumen per steradian	x 1^	
candle, new	candela	x 1^	
	candlepower per square centimeter	x 60^	
candle, standard	candela	x 1^	
electromagnetic spectrum	See APPENDIX, "electromagnetic spectrum"		
foot-candle	lumen per square foot	x 1^	
	lumens per square meter	x 10.764	*x 11*
	lux	x 10.764	*x 11*
foot-candle, apparent	foot-lambert	x 1^	
foot-lambert	candelas per square foot	x 0.318 31	*x 0.32*
	candelas per square meter	x 3.426 3	*x 3.4*
	lumen per square foot	x 1^	
hefner	10-candlepower pentane candle	x 0.090	
	candela	x 0.92	*x 0.9*
	candle, international	x 0.90	
	carcel	x 0.094	
	English candle	x 0.864	*x 0.9*
illuminance	lux	x 1^	
lambert	candelas per square centimeter	/pi or x 0.318 31	*x 0.32*
	candelas per square foot	x 295.72	*x 300*
	candelas per square inch	x 2.053 6	*x 2*
	candelas per square meter	x 10 000/pi or x 3 183.1	*x 3 200*
	lumen per square centimeter	x 1^	
lumen	spherical candle power	/4 pi or x 0.079 577	*x 0.08*
lumen per square foot	foot-candle	x 1^	
	lumens per square meter	x 10.764	*x 11*
lumen per square meter	lumens per square foot	x 0.092 903	*x 0.09*
	lux	x 1^	
	phots	/10 000^ or x 1^E-04	
luminance	candela per square meter	x 1^	
luminous flux	lumen	x 1^	
lux	lumen per square meter	x 1^	
	foot-candles	x 0.092 903	*x 0.09*
magnitude decrease of one (for relative brightness of celestial body)	apparent brightness increase	x 100^ to power 0.2 or x 2.511 9	*x 2.5*
nit	stilb	x 1^	

TABLE 3
Measurement Conversions, By Group

All measurement units are US, unless otherwise noted. All number denominations "billion" and higher are US, unless otherwise noted.

CONVERT	TO EQUIVALENT	BY PRECISELY	OR WITHIN ± 5.0 %
phot	lumen per square centimeter	x 1^	
	lumens per square meter	x 10 000^	
	lux	x 10 000^	
specific refractivity	ratio of refractivity of a medium to its density	x 1^	
spherical candlepower	lumens	x 4 pi or x 12.566	*x 13*
stilb	candelas per square centimeter	x 1^	
LINEAR ACCELERATION			
centimeter per second per second	inches per second per second	x 0.393 70	*x 0.4*
foot per second per second	centimeters per second per second	x 30.48^	*x 30*
	meters per second per second	x 0.304 8^	*x 0.3*
g	See "gravity, acceleration of"	x 1^	
gal (for geodesy)	centimeters per second per second	x 1^	
	meters per second per second	/100^ or x 0.01^	
gravity, acceleration of[8]	centimeters per second per second	x 980.665^	*x 1 000*
	feet per second per second	x 32.174 0	*x 32*
	meters per second per second	x 9.806 65^	*x 10*
inch per second per second	centimeters per second per second	x 2.54^	*x 2.5*
	meters per second per second	x 0.025 4^	*/40*
kilometer per hour per second	centimeters per second per second	x 250/9^ or x 27.778	*x 28*
	feet per second per second	x 0.911 34	*x 0.9*
	meters per second per second	x 5/18^ or x 0.277 78	*x 0.28*
meter per second per second	feet per second per second	x 3.280 8	*x 3.3*
	kilometers per hour per second	x 3.6^	
	miles per hour per second	x 2.236 9	*x 2.2*
mile per hour per second	feet per second per second	x 22/15^ or x 1.466 7	*x 1.5*
	kilometers per hour per second	x 1.609 3	*x 1.6*
	meters per second per second	x 0.447 04	*x 0.45*
LINEAR DISTANCE (see LENGTH)			
LINEAR SPEED			
Beaufort scale (for winds)	See APPENDIX, "wind speeds"		
flank speed (for ships)	maximum-capable speed of ship	x 1	
foot per hour	meters per second	x 8.466 7 E-05	*x 8.5 E-05*
foot per minute	meters per second	x 0.005 08^	*x 0.005*
foot per second	meters per second	x 0.304 8^	*x 0.3*
FPP tornado scale	See APPENDIX, "wind speeds"		
Fujita intensity scale (for winds)	See APPENDIX, "wind speeds"		
hypersonic	Mach above 5	x 1	
inch per second	meters per second	x 0.025 4^	*x 0.025*
kilometer per hour	centimeters per second	x 27.778	*x 28*
	feet per minute	x 54.681	*x 55*
	feet per second	x 0.911 34	*x 0.9*
	knots (intl.)	x 250/463^ or x 0.539 957	*x 0.54*

[8] Standard conditions for the acceleration of gravity are: 9.806 65 meters per second per second = 32.174 0 feet per second per second, at sea level and latitude 45 degrees. At other locations, the acceleration may differ within a span of more than 0.5 percent of the standard value.

TABLE 3
Measurement Conversions, By Group

All measurement units are US, unless otherwise noted. All number denominations "billion" and higher are US, unless otherwise noted.

CONVERT	TO EQUIVALENT	BY PRECISELY	OR WITHIN ± 5.0 %
	meter per second	x 5/18^ or x 0.277 78	*x 0.28*
	meters per minute	x 50/3^ or x 16.667	*x 17*
	miles per hour	x 0.621 37	*x 0.6*
knot	miles, statute per hour	x 1.150 8	*x 1.2*
knot (Brit.)	knot (intl.)	x 1^	
knot (intl.)	feet per hour	x 6 945/1.143^ or x 6 076.1	*x 6 000*
	feet per second	x 1.687 8	*x 1.7*
	kilometers per hour	x 1.852^	*x 1.9*
	meters per second	x 463/900^ or x 0.514 44	*x 0.5*
	miles per hour (intl.)	x 1.150 8	*x 1.2*
	mile, nautical per hour (intl.)	x 1^	
	yards per hour	x 2 025.4	*x 2 000*
knot (US)	knot (intl.)	x 1^	
knot (US) [21]	mile, nautical (intl.) per hour	x 1^	
M	Mach number	x 1^	
Mach	Mach number	x 1^	
Mach 1 to 5	supersonic speed	x 1	
Mach approximately 1	transonic speed	x 1	
Mach greater than 5	hypersonic speed	x 1	
Mach less than 1	subsonic speed	x 1	
Mach number	ratio of object speed to sound speed in a compressible fluid	x 1^	
meter per minute	centimeters per second	x 5/3^ or x 1.666 7	*x 1.7*
	feet per minute	x 3.280 8	*x 3.3*
	feet per second	x 0.054 681	*x 0.055*
	kilometers per hour	x 0.06^	
	kilometers per minute	/1 000^ or x 0.001^	
	miles per hour	x 0.037 282	*x 0.037*
	miles per minute	x 6.213 7 E-04	*x 6 E-04*
meter per second	feet per minute	x 196.85	*x 200*
	feet per second	x 3.280 8	*x 3.3*
	kilometers per hour	x 3.6^	
	miles per hour	x 2.236 9	*x 2.2*
mile per hour	centimeters per second	x 44.704^	*x 45*
	feet per minute	x 88^	*x 90*
	feet per second	x 22/15^ or x 1.466 7	*x 1.5*
	kilometers per hour	x 1.609 3	*x 1.6*
	knots	x 0.868 98	*x 0.87*
	meters per minute	x 26.822	*x 27*
	meters per second	x 0.447 04^	*x 0.45*
	miles, nautical per hour	x 0.868 98	*x 0.87*
mile per minute	centimeters per second	x 2 682.2	*x 2 700*
	feet per second	x 88^	*x 90*
	kilometers per minute	x 1.609 3	*x 1.6*
	meters per second	x 26.822	*x 27*
	miles per hour	x 60^	
mile per second	meters per second	x 1 609.3	*x 1 600*
mile (intl.) per hour	mile per hour	x 1	
mile, nautical per hour	miles per hour	x 1.150 8	*x 1.2*

[21] A ship's log line for measuring speed is marked in segments that are knotted every 47 feet, 3 inches, which equals 14.402 meters. The ship's speed in nautical miles per hour, or knots, equals almost exactly the number of line segments or knots that are counted in 28 seconds as the line is unreeled.

TABLE 3
Measurement Conversions, By Group

All measurement units are US, unless otherwise noted. All number denominations "billion" and higher are US, unless otherwise noted.

CONVERT	TO EQUIVALENT	BY PRECISELY	OR WITHIN ± 5.0 %
mile, statute per hour	mile per hour	x 1^	
Saffir-Simpson hurricane scale	See APPENDIX, "wind speeds"		
speed of light (in vacuum)	kilometers per second	x 2.997 9 E+05	*x 3 E+05*
	miles per second	x 1.862 8 E+05	
subsonic	Mach less than 1	x 1	
	object speed less than sound speed	x 1	
supersonic	Mach 1 to 5	x 1	
	object speed greater than speed of sound	x 1	
tornado	See APPENDIX, "wind speeds"		
transonic	Mach approximately 1	x 1	
	object speed near speed of sound	x 1	
tropical wind scale	See APPENDIX, "wind speeds"		
wind	See APPENDIX, "wind speeds"		
MAGNETISM			
abampere-turn	ampere-turns	x 10^	
	gilberts	x 4 pi or x 12.566	*x 1.3*
abampere-turn per centimeter	ampere-turns per centimeter	x 10^	
	ampere-turns per inch	x 25.4^	*x 25*
	oersteds	x 4 pi or x 12.566	*x 13*
ampere-turn	abampere-turns	/10^ or x 0.1^	
	gilberts	x 0.4 pi or x 1.256 6	*x 1.3*
ampere-turn per centimeter	abampere-turns per centimeter	/10^ or x 0.1^	
	ampere-turns per inch	x 2.54^	*x 2.5*
	oersted	x 0.4 pi or x 1.256 6	*x 1.3*
ampere-turn per inch	abampere-turns per centimeter	x 5/127^ or x 0.0393 70	*x 0.04*
	ampere-turns per centimeter	x 50/127^ or x 0.393 70	*x 0.4*
	ampere-turns per meter	x 5 000/127^ or x 39.370	*x 40*
	gilberts per centimeter	x 0.494 74	*x 0.5*
ampere-turn per meter	ampere-turns per centimeter	x 0.01^	
	ampere-turns per inch	x 0.025 4^	*x 0.025*
	gilberts per centimeter	x 0.012 566	*x 0.013*
	oersteds	x 0.012 566	*x 0.013*
Bohr magneton	joule-square meters per weber	x 9.28 E-24	*x 9 E-24*
electromagnetic spectrum	See APPENDIX, "electromagnetic spectrum"		
gamma (magnetic)	gauss	x 1^E-05	
	line per square centimeter	x 1^	
	lines per square inch	x 6.451 6^	*x 6.5*
	maxwell per square centimeter	x 1^	
	oersteds	x 1^E-05	
	teslas	x 1^E-09	
	webers per square centimeter	x 1^E-08	
	webers per square inch	x 6.451 6^ E-08	*x 6.5 E-08*
	webers per square meter	x 1^E-04	
gauss (magnetic)	teslas	x 1^E-04	
gilbert	abampere-turns	/4 pi or x 0.079 577	*x 0.08*
	amperes	x 0.795 77	*x 0.8*
	ampere-turns	/0.4 pi or x 0.795 77	*x 0.8*
gilbert per centimeter	ampere-turns per inch	x 2.021 3	*x 2*
	ampere-turns per meter	x 79.577	*x 80*

TABLE 3
Measurement Conversions, By Group

All measurement units are US, unless otherwise noted. All number denominations "billion" and higher are US, unless otherwise noted.

CONVERT	TO EQUIVALENT	BY PRECISELY	OR WITHIN ± 5.0 %
	oersted	x 1^	
line	maxwell	x 1^	
line per square centimeter	gauss	x 1^	
line per square inch	gauss	x 0.155 00	*x 0.16*
	webers per square centimeter	x 1.550 0 E-09	*x 1.6 E-09*
	webers per square inch	x 1^E-08	
	webers per square meter	x 1.550 0 E-05	*x 1.6 E-05*
magnetic field strength	ampere per meter	x 1^	
maxwell	line	x 1^	
	webers	x 1^E-08	
nuclear magneton	Bohr magneton	/1837 or x 5.443 7 E-04	
oersted	abampere-turns per centimeter	/4 pi or x 0.079 577	*x 0.08*
	ampere-turns per centimeter	x 10/4 pi or x 0.795 77	*x 0.8*
	ampere-turns per inch	x 2.021 3	*x 2*
	ampere-turns per meter	x 79.577	*x 80*
	gilbert per centimeter	x 1^	
permeability (magnetic)	1/reluctivity	x 1^	
	henry per meter	x 1^	
permeance	1/reluctance	x 1^	
rel	ampere-turn per maxwell	x 1^	
reluctance	1/permeance	x 1^	
reluctivity	1/permeability (magnetic)		
tesla	gammas	x 1^E+09	
	gauss	x 1^E+04	
	weber per square meter	x 1^	
unit pole	webers	x 1.256 6 E-07	*x 1.3 E-07*
weber	maxwells	x 1^E+08	
	unit poles	x 7.957 7 E+06	*x 8 E+06*
weber per square centimeter	gauss	x 1^E-08	
	lines per square inch	x 6.451 6^	*x 6.5*
weber per square inch	gauss	x 1.550 0 E+07	*x 1.6 E+07*
	webers per square centimeter	x 0.155 00	*x 0.16*
weber per square meter	tesla	x 1^	

MASS and WEIGHT (also see FORCE)

Mass is the substance of a physical body and is a measure of the body's inertia, its resistance to having either its speed or its direction of movement changed. In commercial and everyday usage, the word "weight" is loosely used and usually carries the sense of "mass". More strictly, "weight" means "force", that is, the force on a body caused by the downward pull of gravity. A given mass is constant (if we ignore Einsteinian physics) but its force depends on the local force of gravity, which varies with location. The word "weight" should be replaced by "mass" or "force", according to the application. The SI system unequivocally uses "kilogram" for "mass" and "newton" for "force", whereas the metric system previously used "kilogram" ambiguously for both "mass" and "force".
In this book, the "pound", "kilogram", and their related units are used in the sense of "mass", unless noted as "force". Also, all pounds and related units are "avoirdupois", unless noted as "troy" or "apothecary".

CONVERT	TO EQUIVALENT	BY PRECISELY	OR WITHIN ± 5.0 %
amu	atomic mass unit	x 1^	
atomic mass unit	electron-volts (equivalent energy)	x 9.314 8 E+08	*x 9 E+08*
	ergs (equivalent energy)	x 0.001 492 4	*x 0.001 5*
	grams	x 1.660 5 E-24	*x 1.7 E-24*
	mass of carbon-12 atom	/12^ or x 0.083 333	*x 0.08*
atomic mass, relative	atomic weight	x 1^	
	atomic weight	x 1^	

TABLE 3
Measurement Conversions, By Group

All measurement units are US, unless otherwise noted. All number denominations "billion" and higher are US, unless otherwise noted.

CONVERT	TO EQUIVALENT	BY PRECISELY	OR WITHIN ± 5.0 %
avogram	grams	1/Avogadro number or x 1.660 54 E-24	*x 1.7 E-24*
bag (of cement)	pounds, net	x 94^	*x 90*
bale (of cotton)	pounds	x 500^	
barrel (of cement)	pounds, net	x 376^	*x 380*
barrel (of flour)	pounds	x 196^	*x 200*
barrel (of lime), standard large	pounds, net	x 280^	
barrel (of lime), standard small	pounds, net	x 180^	
bundle (for papermaking)	pounds	x 50	
bundle (for shipping paper)	pounds	x 125	*x 130*
capacity, deadweight	tonnage, deadweight	x 1^	
carat grain (for pearls)	carats	/4^ or x 0.25^	
carat (for gemstones)	grains	x 3.086 5	*x 3*
	ounces, apothecary	x 0.006 430 1	*x 0.006 4*
	ounces, avoirdupois	x 0.007 054 8	*x 0.007*
	ounces, troy	x 0.006 430 1	*x 0.006 4*
	points (jeweler's)	x 100^	
carat (for gold)	See "karat"		
carat (international)	milligrams	x 200^	
carat (US before 1913)	milligrams	x 205.3	*x 200*
carat (US since 1913)	milligrams	x 200^	
carat, metric (for precious metals)	carat (intl.)	x 1^	
	grains	x 3.086 5	*x 3*
cental (Brit.)	hundredweights, short (US)	x 1^	
	pounds, avoirdupois	x 100^	
cental (Can.)	cental (Brit.)	x 1^	
centner	kilograms	x 50^	
centner (for assaying)	dram	x 1^	
centner, double	centner, metric	x 1^	
centner, metric	kilograms	x 100^	
chaldron	Also see VOLUME		
chaldron (for coal, Brit.)	hundredweights (Brit.)	x 25.5	*x 26*
clove (Eng.)	pounds	x 8	
dalton	atomic mass unit	x 1^	
drachm (Brit.)	ounce, apothecary	/8^ or x 0.125^	*x 0.13*
drachm, apothecary (Brit.)	dram, apothecary (US)	x 1^	
dram, apothecary	drams, avoirdupois	x 384/175^ or x 2.194 3	*x 2.2*
	grains	x 60^	
	grams	x 3.887 9	*x 4*
	kilograms	x 0.003 887 9	*x 0.004*
	milligrams	x 3 887.9	*x 4 000*
	ounces, apothecary	/8^ or x 0.125^	*x 0.13*
	ounces, avoirdupois	x 24/175^ or x 0.137 14	*x 0.14*
	ounces, troy	/8^ or x 0.125^	*x 0.13*
	pennyweights	x 2.5^	
	pounds, advoirdupois	x 3/350^ or x 0.008 571 4	*x 0.008 6*
	pounds, apothecary	/96^ or x 0.010 417	*x 0.01*
	pounds, troy	/96^ or x 0.010 417	*x 0.01*
	scruples, apothecary	x 3^	

TABLE 3
Measurement Conversions, By Group

All measurement units are US, unless otherwise noted. All number denominations "billion" and higher are US, unless otherwise noted.

CONVERT	TO EQUIVALENT	BY PRECISELY	OR WITHIN ± 5.0 %
dram, avoirdupois	drams, apothecary	x 175/384^ or x 0.455 73	*x 0.46*
	grains	x 875/32^ or x 27.344	*x 27*
	grams	x 1.771 8	*x 1.8*
	kilograms	x 0.001 771 8	*x 0.001 8*
	milligrams	x 1 771.8	*x 1 800*
	ounces, apothecary	x 175/3 072^ or x 0.056 966	*x 0.057*
	ounces, avoirdupois	/16^ or x 0.062 5^	*x 0.063*
	ounces, troy	x 175/3 072^ or x 0.056 966	*x 0.057*
	pennyweights	x 875/768^ or x 1.139 3	*x 1.1*
	pounds, apothecary	x 175/36 864^ or x 0.004 747 2	*x 0.004 7*
	pounds, avoirdupois	/256^ or x 0.003 906 3	*x 0.004*
	pounds, troy	x 175/36 864^ or x 0.004 747 2	*x 0.004 7*
	scruples, apothecary	x 175/128^ or x 1.367 2	*x 1.4*
dram, avoirdupois (Brit.)	dram, avoirdupois (US)	x 1^	
dram, avoirdupois (Can.)	dram, avoirdupois (US)	x 1^	
firkin	Also see VOLUME		
firkin (of butter) (Brit.)	pounds	x 56^	
flask (of mercury)	pounds	x 76^	
gamma (mass)	grams	x 1^E-06	
	microgram	x 1^	
geepound	slug	x 1^	
grain	carats	x 0.323 99	*x 0.32*
	drams, apothecary	/60^ or x 0.016 667	*x 0.017*
	drams, avoirdupois	x 0.036 571	*x 0.037*
	grams	x 0.064 799	*x 0.065*
	kilograms	x 6.479 9 E-05	*x 6.5 E-05*
	milligrams	x 64.799	*x 65*
	ounces, apothecary	/480^ or x 0.002 083 3	*x 0.002*
	ounces, avoirdupois	2/875^ or x 0.002 285 7	*x 0.002 3*
	ounces, troy	/480^ or x 0.002 083 3	*x 0.002*
	pennyweights	/24^ or x 0.041 667	*x 0.04*
	pounds, apothecary	/5 760^ or x 1.736 1 E-04	*x 1.7 E-04*
	pounds, avoirdupois	/7 000^ or x 1.428 6 E-04	*x 1.4 E-04*
	pounds, troy	/5 760^ or x 1.736 1 E-04	*x 1.7 E-04*
	scruples, apothecary	/20^ or x 0.05^	
grain (Can.)	grain (US)	x 1^	
grain, apothecary	grain, avoirdupois	x 1^	
	grain, troy	x 1^	
grain, apothecary (Brit.)	grain, apothecary (US)	x 1^	
grain, avoirdupois	grain, apothecary	x 1^	
	grain, troy	x 1^	
grain, avoirdupois (Brit.)	grain, avoirdupois (US)	x 1^	
grain, carat	See "carat grain"		
grain, pearl	See "pearl grain"		
grain, troy	grain, apothecary	x 1^	
	grain, avoirdupois	x 1^	
grain, troy (Brit.)	grain, troy (US)	x 1^	
gram	drams, apothecary	x 0.257 21	*x 0.26*
	drams, avoirdupois	x 0.564 38	*x 0.56*
	ergs (equivalent energy)	x 8.987 6 E+20	*x 9 E+20*

TABLE 3
Measurement Conversions, By Group

All measurement units are US, unless otherwise noted. All number denominations "billion" and higher are US, unless otherwise noted.

CONVERT	TO EQUIVALENT	BY PRECISELY	OR WITHIN ± 5.0 %
	grains	x 15.432	*x 15*
	joules (equivalent energy)	x 8.987 6 E+13	*x 9 E+13*
	kilograms	/1 000^ or x 0.001^	
	kilowatt-hours (equivalent energy)	x 2.496 5 E+10	
	milligrams	x 1 000^	
	ounces, apothecary	x 0.032 151	*x 0.032*
	ounces, avoirdupois	x 0.035 274	*x 0.035*
	ounces, troy	x 0.032 151	*x 0.032*
	pennyweights	x 0.643 01	*x 0.64*
	pounds, apothecary	x 0.002 679 2	*x 0.002 7*
	pounds, avoirdupois	x 0.002 204 6	*x 0.002 2*
	pounds, troy	x 0.002 679 2	*x 0.002 7*
	scruples, apothecary	x 0.771 62	*x 0.8*
gramme	gram (US)	x 1^	
gram-atom	chemical-element mass, in grams, equal in number to atomic weight	x 1^	
gram-atomic weight	gram-atom	x 1^	
gram-molecular weight	mole	x 1^	
gram-molecule	mole	x 1^	
hundredweight (Brit.)	pounds, avoirdupois	x 112^	*x 110*
	stones (Brit.)	x 8^	
hundredweight, gross	hundredweight, long	x 1^	
hundredweight, long	hundredweight, gross	x 1^	
	kilograms	x 50.802	*x 50*
	pounds	x 112^	*x 110*
hundredweight, long (Can.)	hundredweight, long (US)	x 1^	
hundredweight, long (US)	hundredweight (Brit.)	x 1^	
hundredweight, net	hundredweight, short	x 1^	
hundredweight, short	hundredweight, net	x 1^	
	kilograms	x 45.359	*x 45*
	ounces, avoirdupois	x 1 600^	
	pounds, avoirdupois	x 100^	
	tons, long	x 0.044 643	*x 0.045*
	tons, metric	x 0.045 359	*x 0.045*
	tons, short	/20^ or x 0.05^	
hundredweight, short (Can.)	hundredweight, short (US)	x 1^	
karat (for gemstones)	See "carat"		
karat (for gold)	See "carat"		
	percent of gold in metal alloy	x 25/6^ or x 4.166 7	*x 4*
kilo (short form)	kilogram	x 1^	
kilogram	carats	x 5 000^	
	drams, apothecary	x 257.21	*x 260*
	drams, avoirdupois	x 564.38	*x 560*
	grains	x 15 432	*x 15 000*
	grams	x 1 000^	
	hundredweights, long	x 0.019 684	*x 0.02*
	hundredweights, short	x 0.022 046	*x 0.022*
	milligrams	x 1^E+06	
	ounces, advoirdupois	x 35.274	*x 35*
	ounces, apothecary	x 32.151	*x 32*

TABLE 3
Measurement Conversions, By Group

All measurement units are US, unless otherwise noted. All number denominations "billion" and higher are US, unless otherwise noted.

CONVERT	TO EQUIVALENT	BY PRECISELY	OR WITHIN ± 5.0 %
	ounces, troy	x 32.151	*x 32*
	pennyweights	x 643.01	*x 640*
	pounds	x 2.204 6	*x 2.2*
	pounds, apothecary	x 2.679 2	*x 2.7*
	pounds, avoirdupois	x 2.204 6	*x 2.2*
	pounds, troy	x 2.679 2	*x 2.7*
	scruples, apothecary	x 771.62	*x 800*
	tons, long	x 9.842 1 E-04	*x 0.001*
	tons, metric	/1 000^ or x 0.001^	
	tons, short	x 0.001 102 3	*x 0.001 1*
kilopond	See FORCE		
mass	weight (force)	/g (see LINEAR ACCELERATION, "g")	
mass unit	atomic mass unit	x 1^	
mass, molar	kilogram per mole	x 1^	
microgram	grains	x 1.543 2 E-05	*x 1.5 E-05*
	grams	x 1^E-06	
millier	kilograms	x 1 000^	
	ton, metric	x 1^	
milligram	carats	/200^ or x 0.005^	
	drams, apothecary	x 2.572 1 E-04	*x 2.5 E-04*
	drams, avoirdupois	x 5.643 8 E-04	*x 5.6 E-04*
	grains	x 0.015 432	*x 0.015*
	grams	/1 000^ or x 0.001^	
	kilograms	/1^E+06 or x 1^E-06	
	ounces, apothecary	x 3.215 1 E-05	*x 3.2 E-05*
	ounces, avoirdupois	x 3.527 4 E-05	*x 3.5 E-05*
	ounces, troy	x 3.215 1 E-05	*x 3.2 E-05*
	pennyweights	x 6.430 1 E-04	*x 6.4 E-04*
	points (jeweler's)	/2^ or x 0.5^	
	pounds, apothecary	x 2.679 2 E-06	*x 2.7 E-06*
	pounds, avoirdupois	x 2.204 6 E-06	*x 2.2 E-06*
	pounds, troy	x 2.679 2 E-06	*x 2.7 E-06*
	scruples, apothecary	x 7.716 2 E-04	*x 8 E-04*
mol	mole	x 1^	
mole (mass)	chemical mass, e.g.,grams or pounds, equal in number to molecular weight	x 1^	
mole (quantity)	See QUANTITY		
molecular mass, relative	molecular weight	x 1^	
molecular weight	sum of atomic weights of all the atoms in a molecule	x 1^	
myriagram (obsolete)	grams	x 10 000^	
neutron rest mass	atomic mass units	x 1.008 7	*x 1*
ounce	ounce, avoirdupois	x 1^	
ounce, apothecary	carats	x 155.52	*x 160*
	drams, apothecary	x 8^	
	drams, avoirdupois	x 3 072/175^ or x 17.554	*x 17*
	grains	x 480^	
	grams	x 31.103	*x 30*
	kilograms	x 0.031 103	*x 0.03*
	milligrams	x 31 103	*x 30 000*
	ounces, avoirdupois	x 1.097 1	*x 1.1*
	pennyweights	x 20^	
	pounds, apothecary	/12^ or x 0.083 333	*x 0.08*

TABLE 3
Measurement Conversions, By Group

All measurement units are US, unless otherwise noted. All number denominations "billion" and higher are US, unless otherwise noted.

CONVERT	TO EQUIVALENT	BY PRECISELY	OR WITHIN ± 5.0 %
	pounds, avoirdupois	x 12/175^ or x 0.068 571	*x 0.07*
	pounds, troy	/12^ or x 0.083 333	*x 0.08*
	scruples, apothecary	x 24^	
ounce, apothecary (Brit.)	ounce, apothecary (US)	x 1^	
ounce, apothecary (Can.)	ounce, apothecary (US)	x 1^	
ounce, avoirdupois	carats	x 141.75	*x 140*
	drams, apothecary	x 175/24^ or x 7.291 7	*x 7*
	drams, avoirdupois	x 16^	
	grains	x 437.5^	*x 440*
	grams	x 28.350	*x 28*
	hundredweights, short	/1 600^ or x 6.25^E-04	*x 6 E-04*
	kilograms	x 0.028 350	*x 0.028*
	milligrams	x 28 350	*x 28 000*
	ounces, apothecary	x 175/192^ or x 0.911 46	*x 0.9*
	ounces, troy	x 175/192^ or x 0.911 46	*x 0.9*
	pennyweights	x 875/48^ or x 18.229	*x 18*
	pounds, apothecary	x 175/2 304^ or x 0.075 955	*x 0.076*
	pounds, avoirdupois	/16^ or x 0.062 5^	*x 0.06*
	pounds, troy	x 175/2 304^ or x 0.075 955	*x 0.08*
	scruples, apothecary	x 175/8^ or x 21.875^	*x 22*
	tons, long	x 2.790 2 E-05	*x 2.8 E-05*
	tons, metric	x 2.83 50 E-05	*x 2.8 E-05*
	tons, short	/32 000^ or x 3.125^E-05	*x 3 E-05*
ounce, avoirdupois (Brit.)	ounce, avoirdupois (US)	x 1^	
ounce, avoirdupois (Can.)	ounce, avoirdupois (US)	x 1^	
ounce, avoirdupois (US)	ounce, avoirdupois (Brit.)	x 1^	
ounce, troy (Can.)	ounce, troy (US)	x 1^	
ounce, troy (for precious metals)	carats	x 155.52	*x 160*
	drams, apothecary	x 8^	
	drams, avoirdupois	x 3 072/175^ or x 17.554	*x 18*
	grains	x 480^	
	grams	x 31.103	*x 30*
	kilograms	x 0.031 103	*x 0.03*
	milligrams	x 31 103	*x 30 000*
	ounces, avoirdupois	x 192/175^ or x 1.097 1	*x 1.1*
	pennyweights	x 20^	
	pounds, apothecary	/12^ or x 0.083 333	*x 0.08*
	pounds, avoirdupois	x 12/175^ or x 0.068 571	
	pounds, troy	/12^ or x 0.083 333	*x 0.08*
	scruples, apothecary	x 24^	
ounce, troy (for precious metals, Brit.)	ounce, troy (US)	x 1^	
pearl grain (for pearls)	carat grain	x 1^	
pennyweight	drams, apothecary	x 0.4^	
	drams, avoirdupois	x 0.877 71	*x 0.9*
	grains	x 24^	
	grams	x 1.555 2	*x 1.6*
	kilograms	x 0.001 555 2	*x 0.001 6*
	milligrams	x 1 555.2	*x 1 600*

TABLE 3
Measurement Conversions, By Group

All measurement units are US, unless otherwise noted. All number denominations "billion" and higher are US, unless otherwise noted.

CONVERT	TO EQUIVALENT	BY PRECISELY	OR WITHIN ± 5.0 %
	ounces, apothecary	/20^ or x 0.05^	
	ounces, avoirdupois	x 48/875^ or x 0.054 857	*x 0.055*
	ounces, troy	/20^ or x 0.05^	
	pounds, apothecary	/240^ or x 0.004 166 7	*x 0.004*
	pounds, avoirdupois	x 3/875^ or x 0.003 428 6	*x 0.003 4*
	pounds, troy	/240^ or x 0.004 166 7	*x 0.004*
	scruples, apothecary	x 1.2^	
pennyweight (Can.)	pennyweight (US)	x 1^	
pennyweight, troy (Brit.)	pennyweight, troy (US)	x 1^	
point (jeweler's)	milligrams	x 2^	
	carat	/100^ or x 0.01^	
pound	pound, avoirdupois	x 1^	
pound, apothecary	drams, apothecary	x 96^	
	drams, avoirdupois	x 36 864/175^ or x 210.65	*x 210*
	grains	x 5 760^	*x 6 000*
	grams	x 373.24	*x 370*
	kilograms	x 0.373 24	*x 0.37*
	milligrams	x 3.732 4 E+05	*x 3.7 E+05*
	ounces, apothecary	x 12^	
	ounces, avoirdupois	x 2 304/175^ or x 13.166	*x 13*
	ounces, troy	x 12^	
	pennyweights	x 240^	
	pounds, avoirdupois	x 144/175^ or x 0.822 86	*x 0.8*
	scruples, apothecary	x 288^	*x 300*
pound, apothecary (Brit.)	pound, apothecary (US)	x 1^	
pound, apothecary (Can.)	pound, apothecary (US)	x 1^	
pound, avoirdupois	drams, apothecary	x 350/3^ or x 116.67	*x 120*
	drams, avoirdupois	x 256^	*x 260*
	grains	x 7 000^	
	grams	x 453.59	*x 450*
	hundredweights, short	/100^ or x 0.01^	
	kilograms	x 0.453 59	*x 11/24 or x 4/9*
	milligrams	x 4.535 9 E+05	*x 4.5 E+05*
	ounces, apothecary	x 175/12^ or x 14.583	*x 15*
	ounces, avoirdupois	x 16^	
	ounces, troy	x 175/12^ or x 14.583	*x 15*
	pennyweights	x 875/3^ or x 291.67	*x 300*
	pound (customary)	x 1^	
	pounds, apothecary	x 175/144^ or x 1.215 3	*x 1.2*
	pounds, troy	x 175/144^ or x 1.215 3	*x 1.2*
	scruples, apothecary	x 350^	
	tons, long	/2 240^ or x 4.464 3 E-04	*x 4.4 E-04*
	tons, metric	x 4.535 9 E-04	*x 4.5 E-04*
	tons, short	/2 000^ or x 5^E-04	
pound, avoirdupois (Brit.)	pound, avoirdupois (US)	x 1^	
pound, avoirdupois (Can.)	pound, avoirdupois (US)	x 1^	
pound, troy	drams, apothecary	x 96^	
	drams, avoirdupois	x 36 864/175^ or x 210.65	*x 210*
	grains	x 5 760^	*x 6 000*

TABLE 3
Measurement Conversions, By Group

All measurement units are US, unless otherwise noted. All number denominations "billion" and higher are US, unless otherwise noted.

CONVERT	TO EQUIVALENT	BY PRECISELY	OR WITHIN ± 5.0 %
	grams	x 373.24	*x 370*
	kilograms	x 0.373 24	*x 0.37*
	ounces, apothecary	x 12^	
	ounces, avoirdupois	x 2 304/175^ or x 13.166	*x 13*
	ounces, troy	x 12^	
	pennyweights	x 240^	
	pounds, avoirdupois	x 144/175^ or x 0.822 86	*x 0.8*
	scruples, apothecary	x 288^	*x 300*
pound, troy (Brit.)	pound, troy (US)	x 1^	
pound, troy (Can.)	pound, troy (US)	x 1^	
pound-atom	mass of element, in pounds, equal in number to atomic weight	x 1^	
pound-molecule	mass of molecule, in pounds, equal in number to molecular weight	x 1^	
proton rest mass	atomic mass units	x 1.007 3	*x 1*
quarter (Brit.)	pounds, avoirdupois	x 28^	
	stones (Brit.)	x 2^	
quartern (for a loaf of bread, Brit.)	pounds	x 4	
quintal (Brit.)	pounds, avoirdupois	x 112^	*x 110*
quintal, long	hundredweight, long	x 1^	
quintal, metric	kilograms	x 100^	
	pounds, avoirdupois	x 220.46	*x 220*
quintal, short	hundredweight, short	x 1^	
scruple (Brit.)	drachms (Brit.)	/3^ or x 0.333 33	*x 0.33*
scruple, apothecary	drams	/3^ or x 0.333 33	*x 0.33*
	drams, avoirdupois	x 128/175^ or x 0.731 43	*x 0.7*
	grains	x 20^	
	grams	x 1.296 0	*x 1.3*
	kilograms	x 0.001 296 0	*x 0.001 3*
	milligrams	x 1 296.0	*x 1 300*
	ounces, apothecary	/24^ or x 0.041 667	*x 0.04*
	ounces, avoirdupois	x 8/175^ or x 0.045 714	*x 0.046*
	ounces, troy	/24^ or x 0.041 667	*x 0.04*
	pennyweights	5/6^ or x 0.833 33	*x 0.8*
	pounds, apothecary	/288^ or x 0.003 472 2	*x 0.003 5*
	pounds, avoirdupois	/350^ or x 0.002 857 1	*x 0.002 9*
	pounds, troy	/288^ or x 0.003 472 2	*x 0.003 5*
scruple, apothecary (Brit.)	scruple, apothecary (US)	x 1^	
slug	kilograms	x 14.594	*x 15*
	pound (force) per foot per second per second	x 1^	
slug, metric	kilogram (force) per meter per second per second	x 1^	
specific weight	kilogram (force) per cubic meter	x 1^	
	pound (force) per cubic foot	x 1^	
	ratio of mass of a substance to its volume	x 1	
stone (Brit.)	pounds, avoirdupois	x 14^	
stone (obsolete, Brit.)	pounds, avoirdupois	x 16	
ton (Brit.)	hundredweights (Brit.)	x 20^	
	pounds, avoirdupois	x 2 240^	*x 2 200*
	stones (Brit.)	x 160^	
	tons, net (US)	x 1.12^	*x 1.1*
	tons, short (US)	x 1.12^	*x 1.1*

TABLE 3
Measurement Conversions, By Group

All measurement units are US, unless otherwise noted. All number denominations "billion" and higher are US, unless otherwise noted.

CONVERT	TO EQUIVALENT	BY PRECISELY	OR WITHIN ± 5.0 %
ton (customary)	ton, short (US)	x 1^	
ton (mass)	Also see ENERGY; FORCE; and VOLUME		
tonnage	Also see VOLUME		
tonnage, cargo	number of long tons (of cargo)	x 1	
	number of metric tons (of cargo)	x 1	
tonnage, deadweight	number of long tons (of fuel, passengers, and cargo fully loaded)	x 1	
	number of metric tons (of fuel, passengers, and cargo fully loaded)	x 1	
tonnage, displacement	See "tonnage, cargo" and "tonnage, deadweight"		
tonnage, gross	See VOLUME		
tonnage, net	See VOLUME		
tonnage, register under-deck	See VOLUME		
tonnage, vessel	See VOLUME		
tonne	ton, metric	x 1^	
ton, assay	grams	x 175/6^ or x 29.167	*x 30*
	milligrams	x 87 500/3^ or x 29 167^	*x 30 000*
ton, gross	hundredweights, gross	x 20^	
	kilograms	x 1 016.0	*x 1 000*
	ton, long	x 1^	
ton, long	hundredweights, long	x 20^	
	hundredweights, short	x 22.4^	*x 22*
	kilograms	x 1 016.0	*x 1 000*
	ounces, avoirdupois	x 35 840^	*x 36 000*
	pounds	x 2 240^	*x 2 200*
	tons, metric	x 1.016 0	*x 1*
	tons, short	x 1.12^	*x 1.1*
	ton, gross	x 1^	
ton, long (Can.)	ton, long (US)	x 1^	
ton, long (of fresh water displaced by ships)	foot, cubic	x 35.9	*x 36*
ton, long (of seawater displaced by ships)	foot, cubic	x 35	
ton, long (US)	ton (Brit.)	x 1^	
ton, metric	hundredweights, short	x 22.046	*x 22*
	kilograms	x 1 000^	
	ounces, avoirdupois	x 35 274	*x 35 000*
	pounds	x 2 204.6	*x 2 200*
	tons, long	x 0.984 21	*x 1*
	tons, short	x 1.102 3	*x 1.1*
ton, net	kilograms	x 907.18	*x 900*
	pounds	x 2 000^	
	ton, short	x 1^	
ton, shipper's	pounds	x 2 240^	*x 2 200*
ton, short	hundredweights, net	x 20^	
	hundredweights, short	x 20^	
	kilograms	x 907.18	*x 900*
	ounces, avoirdupois	x 32 000^	
	pounds	x 2 000^	
	tons, long	x 0.892 86	*x 0.9*
	tons, metric	x 0.907 18	*x 0.9*
	ton, net	x 1^	

TABLE 3
Measurement Conversions, By Group

All measurement units are US, unless otherwise noted. All number denominations "billion" and higher are US, unless otherwise noted.

CONVERT	TO EQUIVALENT	BY PRECISELY	OR WITHIN ± 5.0 %
ton, short (Can.)	ton, short (US)	x 1^	
weight	mass	x g (see LINEAR ACCELERATION, "g")	
	See ENERGY, "force of gravity"		
weight, basis (for paper)	weight of a ream of paper	x 1^	
MASS FLOW (see FLOW RATE)			
MASS PER UNIT AREA (see PRESSURE)			
MASS PER UNIT CAPACITY (see DENSITY)			
MASS PER UNIT LENGTH			
denier (for thread thickness)	grams per meter	/9 000^ or x 1.111 1 E-04	*x 1.1 E-04*
denier (for yarns)	milligrams per meter	/9^ or x 0.111 11	*x 0.11*
gram per meter	ounces per foot	x 0.010 752	*x 0.01*
pound per inch	grams per meter	x 17 858	*x 18 000*
tex (for yarns)	grams per meter	/1 000^ or x 0.001^	
tex (for yarn)	milligram per meter	x 1^	
MASS PER UNIT TIME (see FLOW RATE)			
MASS PER UNIT VOLUME (see DENSITY)			
MOMENT OF INERTIA			
gram-square centimeter	kilogram-square meters	x 1^E-07	
	pound-square feet	x 2.373 0 E-06	*x 2.4 E-06*
	pound-square inches	x 3.417 2 E-04	*x 3.4 E-04*
	slug-square feet	x 7.375 6 E-08	*x 7.4 E-08*
kilogram-square meter	gram-square centimeters	x 1^E+07	
	pound-square feet	x 23.730	*x 24*
	pound-square inches	x 3 417.2	*x 3 400*
	slug-square feet	x 0.737 56	*x 0.74*
pound-square foot	gram-square centimeters	x 4.214 0 E+05	*x 4.2 E+05*
	kilogram-square meters	x 0.042 140	*x 0.042*
	pound-square inches	x 144^	
	slug-square feet	x 0.031 081	*x 0.03*
pound-square inch	gram-square centimeters	x 2 926.4	*x 3 000*
	kilogram-square meters	x 2.926 4 E-04	*x 3 E-04*
	pound-square feet	/144^ or x 0.006 944 4	*x 0.007*
	slug-square feet	x 2.158 4 E-04	*x 2.2 E-04*
slug-square foot	gram-square centimeters	x 1.355 8 E+07	*x 1.4 E+07*
	kilogram-square meters	x 1.355 8	*x 1.4*
	pound-square feet	x 32.174	*x 32*
	pound-square inches	x 4 633.1	*x 4 600*
NUMBER (also see QUANTITY)			
absolute value of a number	magnitude of the number regardless of algebraic sign	x 1^	
aught	zero	x 1^	
Avogadro constant	Avogadro number	x 1^	
Avogadro number	Also see QUANTITY		
	6.022 14 E+23	x 1	*6 E+23*
base number	See "radix"		
base point	See "radix point"		

TABLE 3
Measurement Conversions, By Group

All measurement units are US, unless otherwise noted. All number denominations "billion" and higher are US, unless otherwise noted.

CONVERT	TO EQUIVALENT	BY PRECISELY	OR WITHIN ± 5.0 %
basis point (for loan investment yields)	0.01^ of one percent	x 1^	
billion (Brit.)	trillion (US)	x 1^	
billion (US)	1^E+09	x 1^	
binary	based on the number 2	x 1^	
cent per cent	a hundred for each hundred	x 1^	
centage	percentage	x 1^	
centesimal	divided into hundredths	x 1^	
centillion (Brit.)	1^E+600	x 1^	
centillion (US)	1^E+303	x 1^	
cipher	zero	x 1^	
confidence level (statistical)	See APPENDIX, "numbers"		
decade (number)	ratio of 10 to 1	x 1^	
decillion (Brit.)	novemdecillion (US)	x 1^	
decillion (US)	1^E+33	x 1^	
decimal	based on the number 10	x 1^	
dibit	a binary-number arrangement: 00, 01, 10, or 11	x 1^	
diurnal	daily	x 1^	
duodecillion (Brit.)	1^E+72	x 1^	
duodecillion (US)	1^E+39	x 1^	
duodecimal	based on the number 12	x 1^	
duodenary	based on the number 12	x 1^	
duosexadecimal	duotricinary	x 1^	
duotricinary	based on the number 32	x 1^	
e	2.718 28	x 1	*2.7*
golden ratio	See "ratio, golden"		
golden section	See "ratio, golden"		
googol	10^E+100	x 1^	
googolplex	10^ to power googol	x 1^	
hexadecimal	sexadecimal	x 1^	
infinity	a number without end	x 1	
Julian day number	See TIME		
ln N (N, a number)	log N	x 2.302 59	*x 2.3*
	$\log_e$N	x 1^	
	natural logarithm of N	x 1^	
log N (N, a number)	common logarithm of N	x 1^	
	ln N	x 0.434 294	*x 0.43*
	$\log_{10}$N	x 1^	
logarithm, Briggs	common logarithm	x 1^	
logarithm, common	logarithm to base *10*	x 1^	
logarithm, denary	common logarithm	x 1^	
logarithm, hyperbolic	natural logarithm	x 1^	
logarithm, Napierian	natural logarithm	x 1^	
logarithm, natural	logarithm to base *e*	x 1^	
$\log_{10}$N (N, a number)	common logarithm	x 1^	
	log N	x 1^	
$\log_e$N (N, a number)	ln N	x 1^	
metric prefixes	See Table 1		
milliard (Brit.)	billion (US)	x 1^	
million (Brit.)	million (US)	x 1^	
million (US)	1^E+06	x 1^	
naught	zero	x 1^	

TABLE 3
Measurement Conversions, By Group

All measurement units are US, unless otherwise noted. All number denominations "billion" and higher are US, unless otherwise noted.

CONVERT	TO EQUIVALENT	BY PRECISELY	OR WITHIN ± 5.0 %
nonary	novenary	x 1^	
nonillion (Brit.)	septendecillion (US)	x 1^	
nonillion (US)	1^E+30	x 1^	
nought	zero	x 1^	
novemdecillion (Brit.)	1^E+114	x 1^	
novemdecillion (US)	1^E+60	x 1^	
novenary	based on the number 9	x 1^	
number denominations, relative values of (above 100)	See APPENDIX, "numbers"		
number, decimal	percent	x 100^	
octal	based on the number 8	x 1^	
octillion (Brit.)	quindecillion (US)	x 1^	
octillion (US)	1^E+27	x 1^	
octodecillion (Brit.)	1^E+108	x 1^	
octodecillion (US)	1^E+57	x 1^	
octodenary	based on the number 18	x 1^	
octonary	octal	x 1^	
order of magnitude	See SIZE, "order(s) of magnitude"		
ought	zero	x 1^	
percent	decimal number	/100^ or x 0.01^	
percentage point	the difference between two percentages, e.g., 40% minus 32 % = 8 percentage points	x 1^	
pi	3.141 593	x 1	*355/113 or 22/7 or 3*
point, basis (for loan investment yields)	See "basis point"		
point, percentage	See "percentage point"		
point, radix (for numbers)	See "radix point"		
quadragenary	based on the number 40	x 1^	
quadrillion (Brit.)	septillion (US)	x 1^	
quadrillion (US)	1^E+15	x 1^	
quartern	a fourth	x 1^	
quaterdenary	based on the number 14	x 1^	
quaternary	based on the number 4	x 1^	
quattuordecillion (Brit.)	1^E+84	x 1^	
quattuordecillion (US)	1^E+45	x 1^	
quinary	based on the number 5	x 1^	
quindecillion (Brit.)	1^E+90	x 1^	
quindecillion (US)	1^E+48	x 1^	
quintillion (Brit.)	nonillion (US)	x 1^	
quintillion (US)	1^E+18	x 1^	
radix	the base of a number system	x 1^	
	the number of symbol types in a number system	x 1^	
radix point	the character separating the integer and fraction parts of a number	x 1^	
ratio, golden (for esthetic design)	0.618 03 or 1.618 03 [10]	x 1	*3/5 or 5/3*
relative value of number denominations	See APPENDIX, "numbers"		
rms	root mean square	x 1^	

[10] The golden ratio equals - 0.5 ± the square root of 1.25.

TABLE 3
Measurement Conversions, By Group

All measurement units are US, unless otherwise noted. All number denominations "billion" and higher are US, unless otherwise noted.

CONVERT	TO EQUIVALENT	BY PRECISELY	OR WITHIN ± 5.0 %
Roman numbers	See APPENDIX, "numbers"		
root mean square	square root of arithmetic mean of the squares of a group of numbers	x 1	
root-mean-square deviation	standard deviation	x 1^	
score	20^	x 1^	
semi-infinite	extending to infinity in one direction or dimension	x 1	
senary	based on the number 6	x 1^	
septenary	based on the number 7	x 1^	
septendecillion (Brit.)	1^E+102	x 1^	
septendecillion (US)	1^E+54	x 1^	
septendecimal	based on the number 17	x 1^	
septillion (Brit.)	tredecillion (US)	x 1^	
septillion (US)	1^E+24	x 1^	
sexadecimal	based on the number 16	x 1^	
sexagenary	based on the number 60	x 1^	
sexagesimal	sexagenary	x 1^	
sexdecillion (Brit.)	1^E+96	x 1^	
sexdecillion (US)	1^E+51	x 1^	
sextillion (Brit.)	undecillion (US)	x 1^	
sextillion (US)	1^E+21	x 1^	
sigma (statistical)	standard deviation	x 1^	
solo	one alone	x 1^	
ternary	based on the number 3	x 1^	
tredecillion (Brit.)	1^E+78	x 1^	
tredecillion (US)	1^E+42	x 1^	
tricenary	based on the number 30	x 1^	
trillion (Brit.)	quintillion (US)	x 1^	
trillion (US)	1^E+12	x 1^	
undecillion (Brit.)	1^E+66	x 1^	
undecillion (US)	1^E+36	x 1^	
undecimal	based on the number 11	x 1^	
vicenary	based on the number 20	x 1^	
vigesimal	vicenary	x 1^	
vigintillion (Brit.)	1^E+120	x 1^	
vigintillion (US)	1^E+63	x 1^	
wave number	See LENGTH		
OPTICS			
density (optical)	log of 1/transmittance	x 1	
diopters (optical power of lens)	1/centimeters (of focal length)	x 100^	
	1/meters (of focal length)	x 1^	
f-number	ratio of lens focal length of lens to aperture effective diameter	x 1^	
f-stop	f-number	x 1^	
magnification	ratio of optical-image size to object size	x 1^	
meters (focal length of lens)	1/diopters (of optical power)	x 1^	
power, optical	magnification	x 1^	
transmittance	10 to power of negative density (optical)	x 1	
PERIOD (see TIME)			
PERMEABILITY (magnetic) (see MAGNETISM)			

TABLE 3
Measurement Conversions, By Group

All measurement units are US, unless otherwise noted. All number denominations "billion" and higher are US, unless otherwise noted.

CONVERT	TO EQUIVALENT	BY PRECISELY	OR WITHIN ± 5.0 %
PERMEABILITY (porous flow)			
darcy (for oil fields)	square meters	x 9.869 2 E-13	*x 10 E-13*
	square micrometers	x 0.986 92	*x 1*
perm-inch (@ 0 deg. C, for water vapor)	kilograms per pascal-second-meter	x 1.453 2 E-12	*x 1.5 E-12*
perm-inch (@ 23 deg. C, for water vapor)	kilograms per pascal-second-meter	x 1.459 3 E-12	*x 1.5 E-12*
PERMEANCE			
perm (@ 0 deg. C, for water vapor)	kilograms per pascal-second-square meter	x 5.721 4 E-11	*x 6 E-11*
perm (@ 23 deg. C, for water vapor)	kilograms per pascal-second-square meter	x 5.745 3 E-11	*x 6 E-11*
POWER (also see ELECTRICITY) *The "horsepower" is that of 550 foot-pounds (force) per second, unless otherwise noted.*			
bel	decibels	x 10^	
	log of ratio of two power levels	x 1	
Btu per hour	tons of refrigeration	/12 000 or x 8.333 3 E-05	*x 8 E-05*
Btu per minute	tons of refrigeration	/200^ or x 0.005^	
Btu (IT) per hour	calories (IT) per second	x 4.199 9	*x 4*
	foot-pounds (force) per second	x 0.216 16	*x 0.22*
	horsepower	x 3.930 1 E-04	*x 4 E-04*
	watts	x 0.293 07	*x 0.3*
Btu (IT) per minute	foot-pounds (force) per second	x 12.969	*x 13*
	horsepower	x 0.023 581	*x 0.024*
	kilowatts	x 0.0175 84	*x 0.018*
	watts	x 17.584	*x 18*
Btu (IT) per second	watts	x 1 055.1	*x 1 100*
Btu (thermochemical) per hour	watts	x 0.292 88	*x 0.3*
Btu (thermochemical) per minute	watts	x 17.573	*x 18*
Btu (thermochemical) per second	watts	x 1 054.4	*x 1 100*
calorie (thermochemical) per second	watts	x 4.184^	*x 4*
clusec (vacuum-pumping power)	watts	x 1.333 E-06	*x 1.3 E-06*
decibel (for electric power ratio)	nepers	x 0.115 13	*x 23/200 or /9*
	ratio of two electric power levels	x 10 to power {no. of decibels x 0.1}	
erg per second	watts	x 1^E-07	
foot-pound (force) per hour	watts	x 3.766 2 E-04	*x 3.8 E-04*
foot-pound (force) per minute	watts	x 0.022 597	*x 0.023*
foot-pound (force) per second	horsepower	/550^ or x 0.001 818 2	*x 0.001 8*
	horsepower, metric	x 0.001 843 4	*x 0.001 8*
	kilocalories (IT) per second	x 3.238 3 E-04	*x 3.2 E-04*
	meter-kilograms (force) per second	x 0.138 25	*x 0.14*
	watts	x 1.355 8	*x 1.4*
frigorie (for refrigeration, European)	Btu per minute	x 50 (approximate)	
horsepower	foot-pounds (force) per minute	x 33 000^	

TABLE 3
Measurement Conversions, By Group

All measurement units are US, unless otherwise noted. All number denominations "billion" and higher are US, unless otherwise noted.

CONVERT	TO EQUIVALENT	BY PRECISELY	OR WITHIN ± 5.0 %
	foot-pounds (force) per second	x 550^	
	horsepower, metric	x 1.013 9	*x 1*
	kilocalories (IT) per minute	x 10.686	*x 11*
	kilocalories (IT) per second	x 0.178 11	*x 0.18*
	kilocalories (thermochemical) per minute	x 10.694	*x 11*
	kilocalories (thermochemical) per second	x 0.178 23	*x 0.18*
	kilowatts	x 0.745 70	*x 0.75*
	meter-kilograms (force) per second	x 76.040	*x 76*
	watts	x 745.70	*x 750*
horsepower, boiler	Btu (IT) per hour	x 33 471	*x 33 000*
	horsepower	x 13.155	*x 13*
	kilowatts	x 9.809 5	*x 10*
	square feet of heating surface	x 10^	
	watts	x 9 809.5	*x 10 000*
horsepower, electric	watts	x 746^	*x 750*
horsepower, metric	foot-pounds (force) per second	x 542.48	*x 540*
	horsepower	x 0.986 32	*x 1*
	kilocalories (IT) per second	x 0.175 67	*x 0.18*
	kilowatts	x 0.735 50	*x 0.74*
	meter-kilograms (force) per second	x 75^	
	watts	x 735.50	*x 740*
joule per second	watt	x 1^	
kilocalorie (IT) per minute	foot-pounds (force) per second	x 51.467	*x 50*
	horsepower	x 0.093 577	*x 0.09*
	kilowatts	x 0.069 78^	*x 0.07*
	meter-kilograms (force) per second	x 7.115 6	*x 7*
	watts	x 69.78^	*x 70*
kilocalorie (IT) per second	foot-pounds (force) per second	x 3 088.0	*x 3 000*
	horsepower	x 5.614 6	*x 5.6*
	horsepower, metric	x 5.692 5	*x 5.7*
	kilowatts	x 4.186 8^	*x 4*
	meter-kilograms (force) per second	x 426.93	*x 430*
	watts	x 4 186.8^	*x 4 000*
kilocalorie (thermochemical) per minute	watts	x 10 46/15^ or x 69.733	*x 70*
kilocalorie (thermochemical) per second	foot-pounds (force) per second	x 3 086.0	*x 3 000*
	horsepower	x 5.610 8	*x 5.6*
	horsepower, metric	x 5.688 7	*x 5.7*
	kilowatts	x 4.184^	*x 4*
	meter-kilograms (force) per second	x 426.65	*x 430*
kilowatt	Btu per minute	x 56.869	*x 57*
	foot-pounds (force) per minute	x 44 254	*x 44 000*
	foot-pounds (force) per second	x 737.56	*x 740*
	horsepower	x 1.341 0	*x 1.3*
	horsepower, metric	x 1.359 6	*x 1.4*
	kilocalories (IT) per minute	x 14.331	*x 14*
	kilocalories (IT) per second	x 0.238 85	*x 0.24*
	meter-kilograms (force) per second	x 101.97	*x 100*
meter-kilogram (force) per second	foot-pounds (force) per second	x 7.233 0	*x 7*

TABLE 3
Measurement Conversions, By Group

All measurement units are US, unless otherwise noted. All number denominations "billion" and higher are US, unless otherwise noted.

CONVERT	TO EQUIVALENT	BY PRECISELY	OR WITHIN ± 5.0 %
	horsepower	x 0.013 151	*x 0.013*
	horsepower, metric	/75^ or x 0.013 333	*x 0.013*
	kilocalories (IT) per second	x 0.002 342 3	*x 0.002 3*
	kilowatts	x 0.009 806 6	*x 0.001*
neper (for power)	decibels	x 8.685 9	*x 200/23 or x 9*
	ratio of two power levels	x e (= 2.718 3) to power {no. of nepers x 2}	
poncelet	foot-pounds (force) per second	x 723.30	*x 720*
	horsepower	x 1.315 1	*x 1.3*
	horsepower, metric	x 4/3^ or x 1.333 3	*x 1.3*
	kilocalories (IT) per second	x 0.234 23	*x 0.23*
	kilowatts	x 0.980 66	*x 1*
	meter-kilograms (force) per second	x 100^	
	watts	x 980.66	*x 1 000*
power density	watt per square meter	x 1^	
radiance	watt per square meter-steradian	x 1^	
radiant intensity	watt per steradian	x 1^	
ratio of two electric power levels	decibels	x 10 log of ratio	
	nepers	x 0.5 ln of ratio	
sound power	See SOUND		
ton of refrigeration	Btu (IT) per hour	x 12 000	
	Btu (IT) per minute	x 200	
	the heat absorbed to melt one short ton of ice in 24 hours	x 1	
	watts	x 3 517	*x 3 500*
ton of refrigeration (Brit.)	Btu per minute	x 237.6	*x 240*
transmission unit (for electric power)	neper	x 1^	
var	volt-ampere reactive	x 1^	
watt	Btu (IT) per hour	x 3.412 1	*x 3.4*
	Btu (IT) per second	x 9.478 2 E-04	*x 9 E-04*
	ergs per second	x 1^E+07	
	foot-pounds (force) per hour	x 2 655.2	*x 2 700*
	foot-pounds (force) per minute	x 44.254	*x 44*
	foot-pounds (force) per second	x 0.737 56	*x 0.74*
	horsepower	x 0.001 341 0	*x 0.001 3*
	joule per second	x 1^	
watt, intl.	watts, absolute	x 1.000 2	*x 1*
PRESSURE; VACUUM; and STRESS			
ambient pressure	environmental pressure [11]	x 1^	
atmosphere, standard [12]	bars	x 1.013 3	*x 1*
	centimeters of mercury [13]	x 76.000	*x 76*
	centimeters of water [13]	x 1 033.3	*x 1 000*
	feet of water [13]	x 33.900	*x 34*
	inches of mercury [13]	x 29.921	*x 30*
	kilograms (force) per square centimeter	x 1.033 2	*x 1*
	kilograms (force) per square meter	x 10 332	*x 10 000*

11 Ambient pressure is the environmental pressure surrounding a device and is not necessarily atmospheric pressure, whether standard or local.

12 Standard atmospheric conditions are: acceleration of gravity = 9.806 65 meters per second per second = 32.174 0 feet per second per second; atmospheric pressure = 760.00 centimeters of mercury = 29.921 3 inches of mercury = 14.695 9 pounds (force) per square inch; temperature = 0.0 deg. C = 32.0 deg. F.

13 Unless otherwise noted, liquid-head conversions are based on: a pressure of one standard atmosphere; temperature for mercury = 0.0 deg. C = 32.0 deg. F; temperature for water = 4.0 deg. C = 39.2 deg. F.

TABLE 3
Measurement Conversions, By Group

All measurement units are US, unless otherwise noted. All number denominations "billion" and higher are US, unless otherwise noted.

CONVERT	TO EQUIVALENT	BY PRECISELY	OR WITHIN ± 5.0 %
	kilopascals	x 101.33	*x 100*
	meters of mercury [13]	x 0.760 00	*x 0.76*
	millibars	x 1 013.3	*x 1 000*
	millimeters of mercury [13]	x 760.00	*x 760*
	newtons per square meter	x 1.013 25 E+05	*x 1 E+05*
	pounds (force) per square inch	x 14.696	*x 14.7 or x 15*
	tons, short (force) per square foot	x 1.058 1	*x 1.1*
	tons, short (force) per square inch	x 0.007 348 0	*x 0.007*
	torrs	x 760.00	*x 760*
atmosphere, technical	kilograms (force) per square centimeter	x 1^	
	pascals	x 9.806 7 E+04	*x 10 E+04*
atmospheric pressure	the omnidirectional pressure created by the weight of the atmosphere at any specific location	x 1^	
bar	atmospheres, standard	x 0.986 92	*x 1*
	dynes per square centimeter	x 1^E+06	
	kilograms (force) per square centimeter	x 1.019 7	*x 1*
	kilopascals	x 100^	
	newtons per square meter	x 1^E+05	
	pascals	x 1^E+05	
	pounds (force) per square foot	x 2 088.5	*x 2 000*
	pounds (force) per square inch	x 14.504	*x 15*
barye	dyne per square centimeter	x 1^	
centimeter of mercury[13]	pascals	x 1 333.2	*x 1 300*
	pounds (force) per square foot	x 27.845	*x 28*
	pounds (force) per square inch	x 0.193 37	*x 0.19*
centimeter of water [13]	pascals	x 98.064	*x 100*
dyne per square centimeter	bars	x 1^E-06	
	pascals	/10^ or x 0.1^	
	pounds (force) per square inch	x 1.450 4 E-05	*x 1.5 E-05*
	torrs	x 7.500 6 E-04	*x 7.5 E-04*
foot of water (@ 60 deg. F)	pounds (force) per square inch	x 0.433 09	*x 0.43*
foot of water [13]	pascals	x 2 989.0	*x 3 000*
gram (force) per square centimeter	pascals	x 98.067	*x 100*
	pounds (force) per square foot	x 2.048 2	*x 2*
	pounds (force) per square inch	x 0.014 223	*x 0.014*
hectopascal	millibar	x 1^	
inch of mercury (@ 60 deg. F)	pascals	x 3 376.9	*x 3 400*
	pounds (force) per square inch	x 0.489 77	*x 0.5*
inch of mercury [13]	atmospheres, standard	x 0.033 421	*x 0.033*
	bars	x 0.033 864	*x 0.034*
	kilograms (force) per square centimeter	x 0.034 532	*x 0.035*
	pascals	x 3 386.4	*x 3 400*
inch of water (@ 60 deg. F)	pascals	x 248.84	*x 250*
	pounds (force) per square inch	x 0.036 091	*x 0.036*
inch of water [13]	pascals	x 249.08	*x 250*

[13] Unless otherwise noted, liquid-head conversions are based on: a pressure of one standard atmosphere; temperature for mercury = 0.0 deg. C = 32.0 deg. F; temperature for water = 4.0 deg. C = 39.2 deg. F.

TABLE 3
Measurement Conversions, By Group

All measurement units are US, unless otherwise noted. All number denominations "billion" and higher are US, unless otherwise noted.

CONVERT	TO EQUIVALENT	BY PRECISELY	OR WITHIN ± 5.0 %
kilogram (force) per square centimeter	atmosphere, technical	x 1^	
	pascals	x 98 067	*x 1 E+05*
	pounds (force) per square inch	x 14.223	*x 100/7 or x 14*
kilogram (force) per square meter	pascals	x 9.806 7	*x 10*
	pounds (force) per square inch	x 0.001 422 3	*/700 or x 0.001 4*
kilogram (force) per square millimeter	pascals	x 9.806 7 E+06	*x 10 E+06*
kilopascal	bars	/100^ or x 0.01^	
	pounds (force) per square inch	x 0.145 04	*x 0.15*
kip per square inch	kilopascals	x 6 894.8	*x 7 000*
	pounds (force) per square inch	x 1 000^	
megadyne per square centimeter	bar	x 1^	
microbar	pascals	/10^ or x 0.1^	
micron of mercury (@ 0 deg. C)	millitorrs	/1 000 or x 0.001	
millibar	pascals	x 100^	
	pounds (force) per square inch	x 0.014 504	*x 0.015*
millimeter of mercury (intl.)	torr	x 1.000 0	*x 1*
millimeter of mercury [13]	atmospheres, standard	x 0.001 315 8	*x 0.001 3*
	pascals	x 133.32	*x 130*
	torr	x 1.000 0	*x 1*
newton per square meter	pascal	x 1^	
ounce (force) per square foot	kilograms (force) per square meter	x 0.305 15	*x 0.3*
ounce (force) per square inch	grams (force) per square centimeter	x 4.394 2	*x 4.4*
ounce (force) per square yard	kilograms (force) per square meter	x 0.033 906	*x 0.034*
pascal	atmospheres, standard	x 9.869 2 E-06	*x 10 E-06*
	atmospheres, technical	x 1.019 7 E-05	*x 1 E-05*
	bars	x 1^E-05	
	microbars	x 10^	
	newton per square meter	x 1^	
	pounds (force) per square inch	x 1.450 4 E-04	*x 1.5 E-04*
	torrs	x 0.007 500 6	*x 0.007 5*
pascal absolute	pascals gage	- ambient pressure [11]	
pascal gage	pascals absolute	+ ambient pressure [11]	
pound (force) per square foot	kilograms (force) per square meter	x 4.882 4	*x 5*
	pascals	x 47.880	*x 50*
	pounds (force) per square inch	/144^ or x 0.006 944 4	*x 0.007*
pound (force) per square inch	atmospheres, standard	x 0.068 046	*x 0.07*
	bars	x 0.068 948	*x 0.07*
	dynes per square centimeter	x 68 948	*x 70 000*
	feet of water (@ 60 deg. F)	x 2.309 0	*x 2.3*
	grams (force) per square centimeter	x 70.307	*x 70*

[11] Ambient pressure is the environmental pressure surrounding a device and is not necessarily atmospheric pressure, whether standard or local.

[13] Unless otherwise noted, liquid-head conversions are based on: a pressure of one standard atmosphere; temperature for mercury = 0.0 deg. C = 32.0 deg. F; temperature for water = 4.0 deg. C = 39.2 deg. F.

TABLE 3
Measurement Conversions, By Group

All measurement units are US, unless otherwise noted. All number denominations "billion" and higher are US, unless otherwise noted.

CONVERT	TO EQUIVALENT	BY PRECISELY	OR WITHIN ± 5.0 %
	inches of mercury (@ 60 deg. F)	x 2.041 8	*x 2*
	inches of water (@ 60 deg. F)	x 27.708	*x 28*
	kilograms (force) per square centimeter	x 0.070 307	*x 0.07*
	kilograms (force) per square meter	x 703.07	*x 700*
	kilopascals	x 6.894 8	*x 7*
	kips per square inch	/1 000^ or x 0.001^	
	millibars	x 68.948	*x 70*
	pascals	x 6 894.8	*x 7 000*
	poundals per square foot	x 4 633.1	*x 4 600*
	pounds (force) per square foot	x 144^	
	torrs	x 51.715	*x 50*
pound (force) per square inch absolute	pounds (force) per square inch gage	- ambient pressure [11]	
pound (force) per square inch gage	pounds (force) per square inch absolute	+ ambient pressure [11]	
poundal per square foot	pascals	x 1.488 2	*x 1.5*
	pounds (force) per square inch	x 2.158 4 E-04	*x 2.2 E-04*
psi	pascals	x 6 894.8	*x 7 000*
	pounds (force) per square inch	x 1^	
psia	pound (force) per square inch, absolute	x 1^	
psid	pound (force) per square inch, differential	x 1^	
	psi	x 1^	
psig	pound (force) per square inch, gage	x 1^	
torr	atmospheres, standard	/760^ or x 0.001 315 8	*x 0.001 3*
	dynes per square centimeter	x 1 333.2	*x 1 300*
	millimeters of mercury [13]	x 1	
	pascals	x 133.32	*x 130*
	pounds (force) per square inch	x 0.019 337	*x 0.02*
vacuum	pressure below atmospheric pressure	x 1	
vacuum, coarse	torrs	x 1 to x 760	
vacuum, high	torrs	x 0.001 or less	
vacuum, low	pressure slightly below atmospheric pressure	x 1	
WC	water column	x 1^	
QUANTITY (also see NUMBER and SIZE)			
Avogadro number	Also see NUMBER	x 1^	
	number of atoms per gram-atom	x 6.022 14 E+23	*x 6 E+23*
	number of molecules per gram-molecule	x 6.022 14 E+23	*x 6 E+23*
baron (of meat)	sirloins or loins	x 2^	
bilateral	sides	x 2^	
bipartite	parts	x 2^	
brace	similar things	x 2^	
chiliad (also see TIME)	items	x 1 000^	
count (for fabrics)	number of warp yarns and weft yarns per inch	x 1^	
couple	similar or related things	x 2^	
decade (quantity)	group of ten	x 1^	
decuple	items	x 10^	

[11] Ambient pressure is the environmental pressure surrounding a device and is not necessarily atmospheric pressure, whether standard or local.

[13] Unless otherwise noted, liquid-head conversions are based on: a pressure of one standard atmosphere; temperature for mercury = 0.0 deg. C = 32.0 deg. F; temperature for water = 4.0 deg. C = 39.2 deg. F.

TABLE 3
Measurement Conversions, By Group

All measurement units are US, unless otherwise noted. All number denominations "billion" and higher are US, unless otherwise noted.

CONVERT	TO EQUIVALENT	BY PRECISELY	OR WITHIN ± 5.0 %
dozen	items	x 12^	
dozen, baker's	items	x 13^	
duet	group of two	x 1^	
duo	group of two	x 1^	
dyad	group of two	x 1^	
einstein	mole of photons	x 1^	
gross	dozens	x 12^	
	items	x 144^	
heptad	group of seven	x 1^	
hexad	group of six	x 1^	
mole (mass)	See MASS		
mole (quantity)	elementary-particle type as specified	x 6.022 5 E+23	*x 6 E+23*
nonuple	items	x 9^	
octad	group of eight	x 1^	
octave	group of eight	x 1^	
octave (for poetry)	lines in stanza	x 8^	
octet	group of eight	x 1^	
octonary	group of eight	x 1^	
octuple	items	x 8^	
pair	related items	x 2^	
quad (quantity)	group of four	x 1^	
quadragesimal	group of 40		
quadrilateral	sides	x 4^	
quadripartite	parts	x 4^	
quadruple	items	x 4^	
quarter (quantity)	item	/4^	
quartern	parts	/4 or x 0.25	
quartet	group of four	x 1^	
quaternary	group of four	x 1^	
quinary	group of five	x 5^	
quinate	arranged by fives	x 1^	
quintet	group of five	x 1^	
quintuple	items	x 5^	
quire	reams	/20^	
	sheets of paper	x 25^ (usual) or x 24^	
ream	quires	x 20^	
	sheets of drawing or handmade paper	x 472^	*x 470*
ream, long (usual)	sheets of paper	x 500^	
ream, perfect	ream, printer's	x 1^	
ream, printer's	sheets of paper	x 516^	*x 500*
ream, short	sheets of paper	x 480^	*x 500*
score	items	x 20^	
septenary	group of seven	x 1^	
septet	group of seven	x 1^	
septuple	items	x 7^	
sexpartite	parts	x 6^	
sextet	group of six	x 1^	
sextuple	items	x 6^	
single	one item	x 1^	
solo	one alone	x 1^	
tithe	parts	/10^ or x 0.1^	
triad	group of three	x 1^	
trilateral	sides	x 3^	

TABLE 3
Measurement Conversions, By Group

All measurement units are US, unless otherwise noted. All number denominations "billion" and higher are US, unless otherwise noted.

CONVERT	TO EQUIVALENT	BY PRECISELY	OR WITHIN ± 5.0 %
trinary	group of three	x 1^	
trio	group of three	x 1^	
tripartite	parts	x 3^	
triumvirate	group of three	x 1^	
troika	triumvirate	x 1^	
QUANTITY PER UNIT LENGTH			
elite (typewriter type)	characters per inch	x 12^	
pica (typewriter type)	Also see LENGTH		
	characters per inch	x 10^	
wales per inch (for fabric, e.g., corduroy)	number of ribs or ridges per inch	x 1^	
RADIOACTIVITY			
absorbed dose	gray	x 1^	
absorbed dose rate	gray per second	x 1^	
becquerel	curies	x 2.702 7 E-11	*x 2.7 E-11*
	radionuclide activity	x 1/second	
coulomb per kilogram	roentgens	x 3 876.0	*x 4 000*
curie	becquerels	x 3.7^ E+10	
disintegration per second (activity of radionuclide)	becquerel	x 1^	
dose equivalent	sievert	x 1^	
gray			
	rads (absorbed dose)	x 100^	
megawatt-hour per kilogram	joules per kilogram	x 3.6^E+09	
neutron per kilobarn	neutrons per square meter	x 1^E+25	
rad (absorbed dose of radiation))	ergs per gram (of absorbed radiation energy)	x 100	
	grays	x 0.01^	
rem (roentgen equivalent man)	radiation dose causing human biological damage equivalent to that from one roentgen of 200-kV x-rays	x 1	
	roentgen of 200-kV X-rays	x 1^	
	sieverts	x 0.01^	
rep (roentgen equivalent physical)	energy absorption of 93 ergs per gram in soft tissue	x 1^	
	ergs per gram (of radiation energy absorbed in soft tissue)	x 93	*x 90*
roentgen	coulombs per kilogram	x 2.58 E-04	*x 2.6 E-04*
rutherford	radioactive disintegrations per second	x 1^E+06	
sievert	rems (roentgen equivalent man)	x 100^	
SIZE (also see QUANTITY)			
absolute value of vector	magnitude regardless of direction	x 1^	
centuple	100 times as large	x 1^	
colloidal-particle size range	meters	x 1^E-09 to 1^E-06	
double	twice as large	x 1^	
em (printer's)	ens (printer's)	x 2^	
	inches	x 0.166 04	*x 0.17*
	points (printer's)	x 12^	
en (printer's)	ems (printer's)	/2^	

TABLE 3
Measurement Conversions, By Group

All measurement units are US, unless otherwise noted. All number denominations "billion" and higher are US, unless otherwise noted.

CONVERT	TO EQUIVALENT	BY PRECISELY	OR WITHIN ± 5.0 %
lumber, 2 x 4 nominal	1-5/8 x 3-5/8 inches (dressed)	x 1	
magnification	See OPTICS		
Modified Mercalli intensity scale	See APPENDIX, "earthquakes"		
nonuple	nine times as large	x 1	
octuple	eight times as large	x 1	
order(s) of magnitude	ratio of values of two similar items [14]	x 10^ to power of the no. of orders of magnitude	
particle size of clouds	microns	x 0.1 to x 10	
particle size of dusts	microns	x above 10	
particle size of molecules	microns	x below 0.001	
particle size of smokes	microns	x 0.001 to x 0.1	
quadruple	four times as large	x 1^	
quarter (size)	the whole	/4^ or x 0.25^	
quintuple	five times as large	x 1^	
septuple	seven times as large	x 1^	
sextuple	six times as large	x 1^	
triple	three times as large	x 1^	
SOUND			
audible sound	See "range"		
beat frequency	frequency difference of two slightly different simultaneous tones	x 1^	
bel	See POWER		
cent (sound)	interval between two frequencies having ratio of 1 200th root of 2 (= 1.0006)	x 1^	
dB (with suffix *A*, *B*, *C*, or *D*)	decibels	x 1^	
dBA (most common) [15]	decibels on weighted *A* scale at 40 phons	x 1^	
dBB	decibels on weighted *B* scale at 70 phons	x 1^	
dBC	decibels on weighted *C* scale at 100 phons	x 1^	
dBD (for aircraft jet noise)	decibels on weighted *D* scale	x 1^	
decibel increase of one	sound-intensity increase factor	x cube root of 2 or x 1.259 9	*x 1.25*
decibel increase of three	sound-intensity increase factor	x 2	
decibel (for sound-power-intensity level)	ratio of measured intensity to reference intensity[16]	x 10 to power {no. of decibels x 0.1}	
decibel (for sound-pressure-amplitude level)	ratio of measured pressure to reference pressure[17]	x 10 to power {no. of decibels x 0.05}	
frequency for musical tone *A*, standard (intl.)	hertz	x 440.0 ± 0.5	
frequency ratio for half-tone intervals (for tempered musical scale)	the 12th root of 2 (= 1.059 5)	x 1	
frequency ratio for whole-tone intervals (for tempered musical scale)	the 6th root of 2 (= 1.122 5)	x 1	
human sensitivity	See "range"		

[14] Orders of magnitude vary by a factor of 10. Examples: one order of magnitude = ratio of 10:1; two orders, 100:1; three orders, 1000:1; etc.
[15] The dBA scale of sound-pressure level is weighted to approximate the equal-loudness contour curve of the human ear.
[16] The reference level for sound power is 1 E-16 watts per square centimeter.
[17] The reference level for sound pressure is 2 E-04 dynes per square centimeter

TABLE 3
Measurement Conversions, By Group

All measurement units are US, unless otherwise noted. All number denominations "billion" and higher are US, unless otherwise noted.

CONVERT	TO EQUIVALENT	BY PRECISELY	OR WITHIN ± 5.0 %
infrasound	See "range"		
octave (sound)	interval between two frequencies having ratio of 2^	x 1^	
	number of half-tones	x 12^	
	number of whole tones	x 6^	
phon	sound-pressure decibels of 1 000-hertz tone matching loudness of another sound	x 1^	
pitch increase of half tone	See "frequency ratio . . . "		
pitch increase of whole tone	See "frequency ratio . . . "		
range, audible sound	hertz	x 20 to x 20 000	
range, human sound sensitivity (power, minimum up to pain)	watts per square centimeter	x 1 E-16 to x 0.01	
range, human sound sensitivity (pressure, minimum up to pain)	dynes per square centimeter	x 2 E-04 to x 2 000	
range, infrasound	hertz	x below 20	
range, ultrasound	hertz	x above 20 000	
ratio of measured sound power to reference power[16]	decibels	x 10 log of ratio	
ratio of measured sound pressure to reference pressure[17]	decibels	x 20 log of ratio	
reverberation time	See TIME		
semitone	half-tone	x 1^	
sone	loudness of 1 000-hertz tone at 40 decibels above listener's audibility threshold	x 1^	
sound intensity	See "sound power level" and "sound pressure level"		
sound power intensity	sound power	x 1^	
sound power level	See "decibel"		
sound pressure amplitude	sound pressure	x 1^	
sound pressure level	micropascals	x 20^	
	See "decibel"		
sound-power increase of 0.1 decibel	sound-power intensity increase	x 10 to power 0.01 or x 1.023 3	*x 1.0*
sound-power increase of one decibel	sound-power intensity increase	x 10 to power 0.1 or x 1.258 9	*x 1.3*
sound-power-level reference	watts per square centimeter	x 1^E-16	
sound-pressure increase of 0.1 decibel	sound-pressure intensity increase	x 10 to power 0.005 or x 1.011 6	*x 1.0*
sound-pressure increase of one decibel	sound-pressure intensity increase	x 10 to power 0.05 or x 1.122 0	*x 1.1*
sound-pressure-level reference	dynes per square centimeter (in air)	x 2^E-04	
	microbars	x 2^E-04	
ultrasound	See "range"		

[16] The reference level for sound power is 1 E-16 watts per square centimeter.
[17] The reference level for sound pressure is 2 E-04 dynes per square centimeter.

TABLE 3
Measurement Conversions, By Group

All measurement units are US, unless otherwise noted. All number denominations "billion" and higher are US, unless otherwise noted.

CONVERT	TO EQUIVALENT	BY PRECISELY	OR WITHIN ± 5.0 %
SPECIFIC GRAVITY			
API scale (for petroleum)	specific gravities, in deg. API	x 1^	
Baume scales (for light and heavy liquids)	specific gravities, in deg. Baume	x 1^	
deg. Barkometer (for tanning industry)	specific gravities above or below 1.000	x 1^	
deg. Quevenne (for milk)	specific gravities above 1.000	x 1^	
deg. Twaddell (for heavier-than-water liquids)	specific gravities above 1.000	x 1^	
relative density	specific gravity	x 1^	
Wobbe index	See ENERGY		
SPEED and VELOCITY (see ANGULAR SPEED and LINEAR SPEED)			
Speed, a scalar quantity, depends on only the rate of movement. Velocity, a vector quantity, depends on both the rate and the direction of movement.			
SPRING RATE (see FORCE PER UNIT LENGTH)			
STRESS (see PRESSURE)			
TEMPERATURE			
absolute zero temperature	deg. Rankine	-273.15	
	kelvins	x 0^	
boiling point of water (ITS)	deg. Celsius (@ 101.325 kPa)	x 100^	
	deg. Fahrenheit (@ 14696 psia)	x 212^	*x 210*
color temperature	See APPENDIX, "temperature color scale"		
deg. Celsius	deg. Fahrenheit	x 9/5^ + 32^	*x 1.8 + 32*
	deg. Rankine	x 9/5^ + 491.67	*x 1.8 + 500*
	deg. Reaumur	x 4/5^ or x 0.8^	
	kelvins	+ 273.15	*+270*
deg. centigrade (obsolete)	deg. Celsius	x 1^	
deg. Fahrenheit	deg. Celsius	- 32^ x 5/9^	*- 32 x 0.56*
	deg. Rankine	+ 459.67	*+ 460*
	deg. Reaumur	- 32^ x 4/9^	*- 32 x 0.44*
	kelvins	+ 459.67 x 5/9^	*+ 460 x 0.56*
deg. Rankine	deg. Celsius	- 491.67 x 5/9^	*- 500 x 0.56*
	deg. Fahrenheit	- 459.67	*- 460*
	kelvins	x 5/9^ or x 0.555 56	*x 0.56*
deg. Reaumur	deg. Celsius	x 5/4^ or x 1.25^	*x 1.3*
	deg. Fahrenheit	x 9/4^ + 32^	*x 2.3 + 32*
freezing point of water (ITS)	deg. Celsius (@ 101.325 kPa)	x 0^	
	deg. Fahrenheit (@ 14.696 psia)	x 32^	
kelvin	deg. Celsius	- 273.15	*- 270*
	deg. Fahrenheit	x 9/5^ - 459.67	*x 1.8 - 460*
	deg. Rankine	x 9/5^ or x 1.8^	
TEMPERATURE SPAN			
deg. Celsius	deg. Fahrenheit	x 9/5^ or x 1.8^	
	deg. Rankine	x 9/5^ or x 1.8^	
	deg. Reaumur	x 4/5^ or x 0.8^	

TABLE 3
Measurement Conversions, By Group

All measurement units are US, unless otherwise noted. All number denominations "billion" and higher are US, unless otherwise noted.

CONVERT	TO EQUIVALENT	BY PRECISELY	OR WITHIN ± 5.0 %
	kelvin	x 1^	
deg. Fahrenheit	deg. Celsius	x 5/9^ or x 0.555 56	*x 0.56*
	deg. Rankine	x 1^	
	deg. Reaumur	x 4/9^ or x 0.444 44	*x 0.44*
	kelvins	x 5/9^ or x 0.555 56	*x 0.56*
deg. Rankine	deg. Celsius	x 5/9^ or x 0.555 56	*x 0.56*
	deg. Fahrenheit	x 1^	
	kelvins	x 5/9^ or 0.555 56	*x 0.56*
deg. Reaumur	deg. Celsius	x 5/4^ or x 1.25^	*x 1.3*
	deg. Fahrenheit	x 9/4^ or x 2.25^	*x 2.3*
kelvin	deg. Celsius	x 1^	
	deg. Fahrenheit	x 9/5^ or x 1.8^	
	deg. Rankine	x 9/5^ or x 1.8^	
THERMAL CONDUCTANCE			
Btu (IT) per day-square foot-deg. F	Btu (IT) per hour-square foot-deg. F	/24^ or x 0.041 667	*x 0.04*
	calories (IT) per hour-square centimeter-deg. C	x 0.020 343	*x 0.02*
	calories (IT) per second-square centimeter-deg. C	x 5.651 0 E-06	*x 5.7 E-06*
	watts per square centimeter-deg. C	x 2.365 9 E-05	*x 2.4 E-05*
Btu (IT) per hour-square foot-deg. F	Btu (IT) per day-square foot-deg. F	x 24^	
	calories (IT) per hour-square centimeter-deg. C	x 0.488 24	*x 0.5*
	calories (IT) per second-square centimeter-deg. C	x 1.356 2 E-04	*x 1.4 E-04*
	watts per square centimeter-deg. C	x 5.678 3 E-04	*x 5.7 E-04*
calorie (IT) per hour-square centimeter-deg. C	Btu (IT) per day-square foot-deg. F	x 49.156	*x 50*
	Btu (IT) per hour-square foot-deg. F	x 2.048 2	*x 2*
	calories (IT) per second-square centimeter-deg. C	/3 600^ or x 2.777 8 E-04	*x 2.8 E-04*
	watts per square centimeter-deg. C	x 0.001 163 0	*x 0.001 2*
calorie (IT) per second-square centimeter-deg. C	Btu (IT) per day-square foot-deg. F	x 1.769 6 E+05	*x 1.8 E+05*
	Btu (IT) per hour-square foot-deg. F	x 7 373.4	*x 7 400*
	calories (IT) per hour-square centimeter-deg. C	x 3 600^	
	watts per square centimeter-deg. C	x 4.186 8^	*x 4*
coefficient of heat transfer	thermal conductance	x 1^	
R-value	x 1/thermal conductance	x 1^	
thermal insulance	x 1/thermal conductance	x 1^	
U-value	thermal conductance	x 1^	
watt per square centimeter-deg. C	Btu (IT) per day-square foot-deg. F	x 42 266	*x 42 000*
	Btu (IT) per hour-square foot-deg. F	x 1 761.1	*x 1 800*
	calories (IT) per hour-square centimeter-deg. C	x 859.85	*x 900*
	calories (IT) per second-square centimeter-deg. C	x 0.238 85	*x 0.24*

TABLE 3
Measurement Conversions, By Group

All measurement units are US, unless otherwise noted. All number denominations "billion" and higher are US, unless otherwise noted.

CONVERT	TO EQUIVALENT	BY PRECISELY	OR WITHIN ± 5.0 %
THERMAL CONDUCTIVITY			
Btu (IT) per day-square foot-deg. F/inch	Btu (IT) per hour-square foot-deg. F/foot	/288^ or x 0.003 472 2	*x 0.003 5*
	calories (IT) per hour-square centimeter-deg. C/centimeter	x 0.051 672	*x 0.05*
	calories (IT) per second-square centimeter-deg. C/centimeter	x 1.435 4 E-05	*x 1.4 E-05*
	watts per square centimeter-deg. C/centimeter	x 6.009 5 E-05	*x 6 E-05*
Btu (IT) per hour-square foot-deg. F/foot	Btu (IT) per day-square foot-deg. F/inch	x 288^	*x 300*
	calories (IT) per hour-square centimeter-deg. C/centimeter	x 14.882	*x 15*
	calories (IT) per second-square centimeter-deg. C/centimeter	x 0.004 133 8	*x 0.004*
	watts per square centimeter-deg. C/centimeter	x 0.017 317	*x 0.017*
calorie (IT) per hour-square centimeter-deg. C/centimeter	Btu (IT) per day-square foot-deg. F/inch	x 19.353	*x 20*
	Btu (IT) per hour-square foot-deg. F/foot	x 0.067 197	*x 0.07*
	calories (IT) per second-square centimeter-deg. C/centimeter	/3 600^ or x 2.777 8 E-04	*x 2.8 E-04*
	watts per square centimeter-deg. C/centimeter	x 0.001 163 0	*x 0.001 2*
calorie (IT) per second-square centimeter-deg. C/centimeter	Btu (IT) per day-square foot-deg. F/inch	x 69 670	*x 70 000*
	Btu (IT) per hour-square foot-deg. F/foot	x 241.91	*x 240*
	calories (IT) per hour-square centimeter-deg. C/centimeter	x 3 600^	
	watts per square centimeter-deg. C/centimeter	x 4.186 8^	*x 4*
k-value	thermal conductivity	x 1^	
watt per square centimeter-deg. C/centimeter	Btu (IT) per day-square foot-deg. F/inch	x 16 640	*x 17 000*
	Btu (IT) per hour-square foot-deg. F/foot	x 57.779	*x 60*
	calories (IT) per hour-square centimeter-deg. C/centimeter	x 859.85	*x 900*
	calories (IT) per second-square centimeter-deg. C/centimeter	x 0.238 85	*x 0.24*
THERMAL POWER PER UNIT AREA			
Btu (IT) per day-square foot	Btu (IT) per hour-square foot	/24^ or x 0.041 667	*x 0.04*
	calories (IT) per hour-square centimeter	x 0.011 302	*x 0.011*
	calories (IT) per second-square centimeter	x 3.139 4 E-06	*x 3 E-06*
	watts per square centimeter	x 1.314 4 E-05	*x 1.3 E-05*
Btu (IT) per hour-square foot	Btu (IT) per day-square foot	x 24^	
	calories (IT) per hour-square centimeter	x 0.271 25	*x 0.27*
	calories (IT) per second-square centimeter	x 7.534 6 E-05	*x 7.5 E-05*
	watts per square centimeter	x 3.154 6 E-04	*x 3.2 E-04*
	watts per square meter	x 3.154 6	*x 3.2*
Btu (IT) per minute-square foot	watts per square inch	x 0.122 11	*x 0.12*

TABLE 3
Measurement Conversions, By Group

All measurement units are US, unless otherwise noted. All number denominations "billion" and higher are US, unless otherwise noted.

CONVERT	TO EQUIVALENT	BY PRECISELY	OR WITHIN ± 5.0 %
Btu (IT) per second-square foot	watts per square meter	x 11 357	*x 11 000*
Btu (thermochemical) per hour per square foot	watts per square meter	x 3.152 5	*x 3.2*
Btu (thermochemical) per minute-square foot	watts per square meter	x 189.15	*x 190*
Btu (thermochemical) per second-square foot	watts per square meter	x 11.349	*x 11*
Btu (thermochemical) per second-square inch	watts per square meter	x 1.634 2 E+06	*x 1.6 E+06*
calorie (IT) per hour-square centimeter	Btu (IT) per day-square foot	x 88.481	*x 90*
	Btu (IT) per hour-square foot	x 3.686 7	*x 3.7*
	calories (IT) per second-square centimeter	/3 600^ or x 2.777 8 E-04	*x 2.8 E-04*
	watts per square centimeter	x 0.001 163 0	*x 0.001 2*
calorie (IT) per second-square centimeter	Btu (IT) per day-square foot	x 3.185 3 E+05	*x 3.2 E+05*
	Btu (IT) per hour-square foot	x 13 272	*x 13 000*
	calories (IT) per hour-square centimeter	x 3 600^	
	watts per square centimeter	x 4.186 8^	*x 4*
calorie (thermochemical) per minute-square centimeter	watts per square meter	x 697.33	*x 700*
erg per second-square centimeter	watts per square meter	/1 000^ or x 0.001^	
heat-flux density	watt per square meter	x 1	
watt per square centimeter	Btu (IT) per day-square foot	x 76 080	*x 76 000*
	Btu (IT) per hour-square foot	x 3 170.0	*x 3 200*
	calories (IT) per hour-square centimeter	x 859.85	*x 860*
	calories (IT) per second-square centimeter	x 0.238 85	*x 0.24*
	watts per square meter	x 10 000^	
watt per square inch	watts per square meter	x 1 550.0	*x 1 600*

TIME; FREQUENCY; and PERIOD

The customary time periods—from "second through"year"—are based on mean solar time, unless otherwise noted.

CONVERT	TO EQUIVALENT	BY PRECISELY	OR WITHIN ± 5.0 %
bicentennial	years	x 200^	
biennium	years	x 2^	
billisecond	nanosecond	x 1^	
bissextile	a leap year	x 1^	
calendar, Gregorian (since 1582)[19]	calendar (customary)	x 1^	
calendar, Julian day	See "Julian day calendar"		
calendar, Julian (superseded 1582)	calendar, Gregorian (since 1582)	+10 days initially [22]	
calendar, New Style	calendar, Gregorian (since 1582)	x 1^	
calendar, Old Style	calendar, Julian (superseded 1582)	x 1^	
centenary	years	x 100^	

[19] In the Gregorian calendar, every year whose number is divisible by four is a leap year except for centesimal years that are not exactly divisible by 400. The year 2000 is a leap year, 2100 is not a leap year.

[22] Gregorian calendar dates minus Julian calendar dates equals + 10 days in the period 1582 to 1700; + 11 days, 1700 to 1800; + 12 days, 1800 to 1900; + 13 days, 1900 to 2100. George Washington's birthday is February 22, 1742 (Gregorian), equivalent to February 11, 1732 (Julian).

TABLE 3
Measurement Conversions, By Group

All measurement units are US, unless otherwise noted. All number denominations "billion" and higher are US, unless otherwise noted.

CONVERT	TO EQUIVALENT	BY PRECISELY	OR WITHIN ± 5.0 %
centennial	years	x 100^	
century	years	x 100^	
Chandler wobble (of Earth around axis)	days	x 440 (approx.)	
chiliad (time)	Also see QUANTITY		
	years	x 1 000^	
cycle per second	hertz	x 1^	
cycle, metonic	lunar months	x 235	*x 240*
	years	x 19	
daily	occurring every day	x 1^	
day	day, calendar	x 1^	
day (customary)	day, mean solar (midnight to midnight)	x 1^	
	hours	x 24^	
	minutes	x 1 440^	*x 1 400*
	seconds	x 86 400^	*x 86 000*
daylight-saving time	See "time, daylight-saving"		
day, calendar	day, mean solar	x 1^	
	hours, mean solar	x 24^	
	minutes, mean solar	x 1 440^	*x 1 400*
	seconds, mean solar	x 86 400^	*x 86 000*
day, civil	day	x 1^	
	hours	x 24^	
day, lunar	24 hr. 50 min. sidereal time	x 1	
day, mean equinoctial	day, mean sidereal	x 1^	
day, mean solar	24 hr. 3 min. 56.56 sec. mean solar time	x 1	
	day (customary)	x 1^	
	days, sidereal	x 1.002 7	*x 1*
	day, calendar	x 1^	
	hours	x 24^	
	minutes	x 1 440^	*x 1 400*
	seconds	x 86 400^	*x 86 000*
day, natural	hours, inequal	x 12	
day, sidereal	23 hr. 56 min. 4.09. sec mean solar time	x 1	
	day, mean solar	x 0.997 27	*x 1*
	hours, mean solar	x 23.934	*x 24*
	hours, sidereal	x 24^	
	minutes, mean solar	x 1 436.1	*x 1 400*
	minutes, sidereal	x 1 440^	*x 1 400*
	seconds, mean solar	x 86 164	*x 86 000*
	seconds, sidereal	x 86 400^	*x 86 000*
decade (time)	years	x 10^	
decay time	time for a pulse or quantity to decline to a specified percent of initial value	x 1^	
decennary	years	x 10^	
Earth ice-age cycle	years	x 100 000 (approx.)	
fortnight	weeks	x 2^	
frequency	1/period	x 1^	
frequency for musical tone *A*, standard (intl.)	See SOUND		
frequency ratio (for sound)	See SOUND		
geologic time	See APPENDIX, "time"		
gestation period, human average	days	x 267	

TABLE 3
Measurement Conversions, By Group

All measurement units are US, unless otherwise noted. All number denominations "billion" and higher are US, unless otherwise noted.

CONVERT	TO EQUIVALENT	BY PRECISELY	OR WITHIN ± 5.0 %
gestation period, human range	days	x 250 to x 290	
half-life	time period for the potency or quantity of a substance to be reduced by one half	x 1^	
hertz	1/second	x 1^	
	cycle per second	x 1^	
hour (customary)	hour, mean solar	x 1^	
hour, inequal (for astrology)	natural day or natural night (according to times of sunrise and sunset)	/12^ or x 0.083 333	*x 0.08*
hour, mean solar	days	/24^ or x 0.041 667	*x 0.04*
	hours, sidereal	x 1.002 7	*x 1*
	minutes	x 60^	
	seconds	x 3 600^	
hour, planetary	hour, inequal	x 1^	
hour, sidereal	days, sidereal	/24^ or 0.041 667	*x 0.04*
	mean solar seconds	x 3 590.2	*x 3 600*
	minutes, sidereal	x 60^	
	seconds, sidereal	x 3 600^	
Julian century	days	x 36 525	*x 37 000*
Julian day calendar (for astronomy)	a calendar whose starting date is the year 4713 B.C.		
Julian day number	Julian-day-calendar sequential number for any specific day [18]	x 1	
kayser	wave number	x 1^	
kilocycles per second	cycles per second	x 1 000^	
kondratieff cycle (for the business cycle)	years	x 50 to x 60	
longitude	See ANGLE, PLANE		
lunation	lunar month	x 1^	
megacycles per second	cycles per second	x 1^E+06	
metonic cycle	See "cycle, metonic"		
Milankovitch Earth-axis-tilt cycle	years	x 41 000 (approx.)	
Milankovitch Earth-equinoctial-precession cycle	years	x 28 500 (approx.)	
Milankovitch Earth-orbit-eccentricity cycle	years	x 100 000 (approx.)	
millenium	years	x 1 000^	
millisecond	second	/1 000^ or x 0.001^	
minute (customary)	minute, mean solar	x 1^	
minute, mean solar	hours, mean solar	/60^ or x 0.016 667	*x 0.017*
	hours, sidereal	x 0.016 712	*x 0.017*
	minutes, sidereal	x 1.002 7	*x 1*
	seconds	x 60^	
	seconds, sidereal	x 60.164	*x 60*
minute, sidereal	hours, sidereal	/60^ or x 0.016 667	*x 0.017*
	minutes, mean solar	x 0.997 27	*x 1*
	seconds, mean solar	x 59.836	*x 60*
	seconds, sidereal	x 60^	
month (customary)	days, calendar	x 28^, 29^, 30^, or 31^	
	month, mean solar	x 1^	
month, anomalistic	days (customary)	x 27.555	*x 28*

[18] Julian-day-number example: January 1, 1960 is day number 2 346 934.

TABLE 3
Measurement Conversions, By Group

All measurement units are US, unless otherwise noted. All number denominations "billion" and higher are US, unless otherwise noted.

CONVERT	TO EQUIVALENT	BY PRECISELY	OR WITHIN ± 5.0 %
month, lunar	29 days 12 hr. 44 min. 2.8 sec. mean solar time	x 1	
	days, mean solar	x 29.531	*x 30*
	hours, mean solar	x 708.73	*x 700*
month, mean calendar	days	x 30.417	
month, nodical	27 days 5 hr. 5 min. 35.8 sec. mean solar time	x 1	
	mean solar days	x 27.212	*x 27*
month, sidereal	27 days 7 hr. 43 min. 11.5 sec. mean solar time	x 1	
	mean solar days	x 27.322	*x 27*
month, synodic	lunar month	x 1^	
month, tropical	27 days, 7 hr. 43 min. 4.7 sec. mean solar time	x 1	
	days	x 27.322	*x 27*
	tropical years	/12^ or x 0.083 333	*x 0.08*
moon	lunar month	x 1^	
nanosecond	seconds	x 1^E-09	
night, natural	hours, inequal	x 12	
nutation (Earth-axis oscillation)	years	x 18.6	*x 18*
period	1/frequency	x 1^	
period, draconic	nodical month	x 1^	
period, half	See "half-life"		
quadrennial	years	x 4^	
quadrennium	years	x 4^	
quarter (time)	months	x 3	
	school years	/4^	
quindecennial	years	x 15^	
quinquagesimal	days	x 50^	
quinquennium	years	x 5^	
quintan	occurring every fifth day	x 1^	
quotidian	daily	x 1^	
reverberation time	time for a cut-off sound to decrease to one-millionth of its initial intensity	x 1^	
second (customary)	second, mean solar	x 1^	
second, ephemeris	years, tropical	x 3.168 9 E-08	*x 3.2 E-08*
second, mean solar	days, mean solar	/86 400^ or x 1.157 4 E-05	*x 1.2 E-05*
	days, sidereal	x 1.160 6 E-05	*x 1.2 E-05*
	hours, mean solar	/3 600^ or x 2.777 8 E-04	*x 2.8 E-04*
	minutes, mean solar	/60^ or x 0.016 667	*x 0.017*
	seconds, sidereal	x 1.002 7	*x 1*
second, sidereal	days, mean solar	x 1.154 2 E-05	*x 1.2 E-05*
	days, sidereal	x 1.157 4 E-05	*x 1.2 E-05*
	hours, sidereal	/3 600^ or x 2.777 8 E-04	*x 2.8 E-04*
	minutes, sidereal	/60^ or x 0.016 667	*x 0.017*
	seconds, mean solar	x 0.997 27	*x 1*
semester	school years	/2^	
sesquicentennial	years	x 150^	
shake	seconds	x 1^E-08	
sigma (time)	seconds	/1 000^ or x 0.001^	
svedberg	seconds	x 1 E-13	
terdiurnal	days	/3	
time constant	0.632 12 (= {e - 1}/e)	x 1	

TABLE 3
Measurement Conversions, By Group

All measurement units are US, unless otherwise noted. All number denominations "billion" and higher are US, unless otherwise noted.

CONVERT	TO EQUIVALENT	BY PRECISELY	OR WITHIN ± 5.0 %
time of day	See APPENDIX, "time"		
time zone	degrees of longitude (per Greenwich standard)	x 15	
time, daylight-saving	time, standard	- 1^ hour	
time, decay	See "decay time"		
time, standard	time, daylight-saving	+ 1^ hour	
tricentennial	years	x 300^	
triennium	years	x 3^	
trimester	months	x 3^	
	school years	/3^ or x 0.333 33	*x 0.3*
vicennial	years	x 20^	
year (customary)	months (customary)	x 12^	
	year, calendar	x 1^	
	year, mean solar	x 1^	
year, academic	year, school	x 1^	
year, anomalistic	365 days 6 hr. 13 min. 53.1 sec.	x 1	
	days (customary)	x 365.26	*x 365*
year, astronomical	365 days 5 hr. 48 min. 46 sec.	x 1	
year, banker's	days	x 360^ or x 365^	
year, bissextile	leap year	x 1^	
year, calendar	days (December 31 midnight to midnight 12^ months later)	x 365^ or x 366^	
year, civil	year	x 1^	
year, common	seconds	x 3.153 6^E+07	*x 3.2 E+07*
	weeks	x 52.143	*x 52*
year, common [19]	days	x 365^	
year, great	years	x 25 800	*x 26 000*
year, Gregorian	days	x 365.242 5	*x 365*
year, Julian	days	x 365.25	*x 365*
year, leap	seconds	x 3.162 2^E+07	*x 3.2 E+07*
year, leap [19]	days	x 366^	
year, legal	year	x 1^	
year, lunar	lunar months	x 12^	
year, mean solar	365 days 5 hr. 48 min. 45.5 sec. mean solar time	x 1	
	days	x 365.242 5	*x 365*
	hours	x 8 765.8	*x 9 000*
	minutes	x 5.259 5 E+05	*x 5 E+05*
	seconds	x 3.155 7 E+07	*x 3.2 E+07*
year, Platonic	year, great	x 1^	
year, school	quarters	x 4^	
	semesters	x 2^	
	trimesters	x 3^	
	year, academic	x 1^	
year, sidereal	365 days 6 hr. 9 min, 9.5 sec. mean solar time	x 1	
	days, mean solar	x 365.26	*x 365*
	seconds, mean solar	x 3.155 8 E+07	*x 3.2 E+07*
year, tropical	365 days 5 hr. 48 min. 45.5 sec. mean solar time	x 1	
	seconds	x 3.155 7 E+07	*x 3.2 E+07*
	year, mean solar	x 1^	

[19] In the Gregorian calendar, every year whose number is divisible by four is a leap year except for centesimal years that are not exactly divisible by 400. The year 2000 is a leap year, 2100 is not a leap year.

TABLE 3
Measurement Conversions, By Group

All measurement units are US, unless otherwise noted. All number denominations "billion" and higher are US, unless otherwise noted.

CONVERT	TO EQUIVALENT	BY PRECISELY	OR WITHIN ± 5.0 %
zone time	standard time within a time zone	x 1^	
TORQUE and BENDING MOMENT			
dyne-centimeter	newton-meters	x 1^E-07	
kilogram (force)-meter	newton-meters	x 9.806 7	*x 10*
newton-meter	dyne-centimeters	x 1^E+07^	
	kilogram (force)-meters	x 0.101 97	*x 0.1*
	ounce (force)-inches	x 141.61	*x 140*
	pound (force)-feet	x 0.737 56	*x 0.74*
	pound (force)-inches	x 8.850 7	*x 9*
ounce (force)-inch	newton-meters	x 0.007 061 6	*x 0.007*
pound (force)-foot	newton-meters	x 1.355 8	*x 1.4*
pound (force)-inch	newton-meters	x 0.112 98	*x 0.11*
Richter scale (for earthquakes) increase of 0.1 magnitude [24]	seismic-moment increase	x 10 to power 0.15 or x 1.4	
Richter scale (for earthquakes) increase of one magnitude [24]	seismic-moment increase	x 10 to power 1.5 or x 32	
TORQUE PER UNIT LENGTH and BENDING MOMENT PER UNIT LENGTH			
pound (force)-foot per inch	newton-meters per meter	x 53.379	*x 53*
pound (force)-inch per inch	newton-meters per meter	x 4.448 2	*x 4/9*
VACUUM (see PRESSURE)			
VELOCITY (see SPEED)			
VISCOSITY and FLUIDITY			
centipoise	pascal-seconds	/1 000^ or x 0.001^	
	pound (force)-seconds per square foot	x 2.088 5 E-05	*x 2.1 E-05*
	pounds (mass) per foot-second	x 6.719 7 E-04	*x 7 E-04*
centistoke	square feet per second	x 1.076 4 E-05	*x 1.1 E-05*
	square meters per second	x 1^E-06	
	square millimeters per second	x 1^	
pascal-second	centipoises	x 1 000^	
poise	centipoises	x 100^	
	dyne-second per square centimeter	x 1^	
	pascal-seconds	/10^ or x 0.1^	
	pound (force)-seconds per square foot	x 0.002 088 5	*x 0.002*
	pounds (mass) per foot-second	x 0.067 197	*x 0.07*
poise (viscosity)	1/rhe (fluidity)	x 1^	
pound (force)-second per square foot	centipoises	x 47 880	*x 48 000*
	pascal-seconds	x 47.880	*x 48*
	poises	x 478.80	*x 480*
	pounds (mass) per foot-second	x 32.174	*x 32*
	slug per foot-second	x 1^	
pound (force)-second per square inch	pascal-seconds	x 6 894.8	*x 7 000*
pound (mass) per foot-hour	pascal-seconds	x 4.133 8 E-04	*x 4 E-04*

[24] See ENERGY; LENGTH; and APPENDIX, "earthquakes".

TABLE 3
Measurement Conversions, By Group

All measurement units are US, unless otherwise noted. All number denominations "billion" and higher are US, unless otherwise noted.

CONVERT	TO EQUIVALENT	BY PRECISELY	OR WITHIN ± 5.0 %
pound (mass) per foot-second	centipoises	x 1 488.2	*x 1 500*
	pascal-seconds	x 1.488 2	*x 1.5*
	poises	x 14.882	*x 15*
	pound (force)-seconds per square foot	x 0.031 081	*x 0.03*
poundal-second per square foot	pascal-seconds	x 1.488 2	*x 1.5*
rhe	1/pascal-second (viscosity)	x 1^	
	1/poise (viscosity)	x 1^	
square foot per second	centistokes	x 92 903	*x 90 000*
	square meters per second	x 0.092 903	*x 0.09*
	stokes	x 929.03	*x 900*
square millimeter per second	centistoke	x 1^	
stoke	square feet per second	x 0.001 076 4	*x 0.001 1*
	square meters per second	x 1^E-04	
viscosity, dynamic	pascal-second	x 1^	
viscosity, kinematic	square meter per second	x 1^	

VOLUME and CAPACITY

In this book, all the US units of "gallon" and smaller are for fluid measure except that the "quart" and "pint", may be for dry measure, if so designated. All the British units of "bushel" down through "gill" are for fluid and dry measure; the "fluid ounce", "fluid drachm", "mimim", and "fluid scruple" are for only fluid measure.

CONVERT	TO EQUIVALENT	BY PRECISELY	OR WITHIN ± 5.0 %
acre-foot	cubic feet	x 43 560^	*x 44 000*
	cubic meters	x 1 233.5	*x 1 200*
	cubic yards	x 1 613.3	*x 1 600*
	gallons	x 3.258 5 E+05	*x 3.3 E+05*
baby (for wine, Brit.)	bottles (Brit.)	/8^ or x 0.125^	*x 0.13*
	liters	x 3/32 or x 0.093 75	*x 0.09*
balthazar (for wine, Brit.)	bottles (Brit.)	x 16^	
	liters	x 12	
barrel (Can.)	gallon (Can.)	x 36^	
barrel (for cement)	See MASS		
barrel (for cisterns, in one State)	gallons	x 36^	
barrel (for cranberries)	bushels, struck measure	x 2.709	*x 2.7*
	cubic inches	x 5 826	*x 6 000*
	quarts, dry	x 5 549/64^ or x 86.703	*x 90*
barrel (for fermented liquor, Federal)	gallons	x 31^	*x 30*
barrel (for flour)	See MASS		
barrel (for fruits, vegetables, except cranberries)	bushels, struck measure	x 3.281	*x 3.3*
	cubic inches	x 7 056	*x 7 000*
	quarts, dry	x 105	*x 100*
barrel (for lime)	See MASS		
barrel (for liquids, in four States)	gallons	x 42^	
barrel (for liquids, in many States)	gallons	x 31.5^	*x 32*
barrel (for petroleum)	gallons	x 42^	
barrel (for petroleum) (42 gallons)	cubic meters	x 0.158 99	*x 0.16*
barrel (for proof spirits, Federal)	gallons	x 40^	

TABLE 3
Measurement Conversions, By Group

All measurement units are US, unless otherwise noted. All number denominations "billion" and higher are US, unless otherwise noted.

CONVERT	TO EQUIVALENT	BY PRECISELY	OR WITHIN ± 5.0 %
barrel (for sand, Can.)	liters	x 81.830	*x 80*
barrel (per service and jurisdiction)	gallons	x 31 to x 42	
barrelage	number of barrels	x 1	
barrel, herring (Can.)	liters	x 145.47	
barrel, Imperial (Brit.)	bushels, Imperial (Brit.)	x 4.5^	
	kilderkins (Brit.)	x 2	
barrel, Imperial (for ships' water, Brit.)	gallons, Imperial (Brit.)	x 36^	
board foot	cubic feet (nominal)	/12^ or x 0.083 333	*x 0.08*
	cubic inches (actual)	x 96 (approx.)	*x 100*
	cubic inches (nominal)	x 144	
	liters (nominal)	x 2.359 7	*x 2 400*
bottle (for wine, Brit.)	liters	x 3/4^ or x 0.75^	
bushel (Can.)	bushel, Imperial (Brit.)	x 1^	
bushel (customary)	bushel, struck measure	x 1^	
bushelage	number of bushels	x 1	
bushel, heaped	bushels, struck measure	x 1.277 8 (often x 1 1/4)	*x 1.3*
	cubic feet	x 1.590 1	*x 1.6*
	cubic inches	x 2 747.7	*x 2 700*
bushel, Imperial struck measure (Brit.)	bushels, struck measure (US)	x 1.032 1	*x 1*
	cubic feet	x 1.284 4	*x 1.3*
	cubic inches	x 2 219.4	*x 2.2*
bushel, Imperial (Brit.)	gallons, Imperial (Brit.)	x 8^	
	pecks (Brit.)	x 4^	
bushel, struck measure	bushels, heaped	x 0.782 62	*x 0.8*
	cubic feet	x 1.244 5	*x 1.2*
	cubic inches	x 2 150.4	*x 2 200*
	cubic meters	x 0.035 239	*x 0.035*
	liters	x 35.239	*x 35*
bushel, Winchester struck (Brit.)	cubic inches	x 2 150.4	*x 2 200*
but	See "butt"		
butt	gallons	x 126	*x 130*
	hogsheads	x 2^	
	liters	x 476.96	*x 500*
butt (Brit.)	gallons, Imperial (Brit.)	x 108	*x 100*
	hogsheads (Brit.)	x 2^	
butt (for ship's water, Brit.)	gallons, Imperial (Brit.)	x 110^	
carboy	gallons	x 5 to x 15	
cart, salt (Can.)	liters	x 490.98	*x 500*
cart, tub (Can.)	liters	x 81.830	*x 80*
chaldron	Also see MASS		
	bushels	x 36	
	cubic meters	x 1.268 6	*x 1.3*
chaldron (Eng.)	bushels, Imperial (Eng.)	x 32 to x 72	
chaldron (for coal, Eng.)	bushels, Imperial (Eng.)	x 36	
	hundredweights (Eng.)	x 25.5	*x 26*
cistern (4 x 2.5 x 3 feet, Brit.)	gallons, Imperial (Brit.)	x 186.96	*x 190*
coomb (Eng.)	bushels, Imperial (Eng.)	x 4	

TABLE 3
Measurement Conversions, By Group

All measurement units are US, unless otherwise noted. All number denominations "billion" and higher are US, unless otherwise noted.

CONVERT	TO EQUIVALENT	BY PRECISELY	OR WITHIN ± 5.0 %
cord foot (for stacked wood, 4 x 4 x 1 feet)	cords	/8 or x 0.125	*x 0.13*
	cubic feet	x 16	
cord (for stacked wood, 4 x 4 x 8 feet)	cubic feet	x 128	*x 130*
	cubic meters	x 3.624 6	*x 3.6*
cubic centimeter	milliliter	x 1^	
cubic decimeter	liter	x 1^	
cubic foot	bushels, struck measure	x 0.803 56	*x 0.8*
	cubic inches	x 1 728^	*x 1 700*
	cubic meters	x 0.028 317	*x 0.028*
	cubic yards	/27^ or x 0.037 037	*x 0.037*
	drams, fluid	x 7 660.1	*x 8 000*
	gallons	x 7.480 5	*x 7.5*
	gills	x 239.38	*x 240*
	liters	x 28.317	*x 28*
	milliliters	x 28 317	*x 28 000*
	minims	x 4.596 0 E+05	*x 4.6 E+05*
	ounces, fluid	x 957.51	*x 1 000*
	pecks	x 3.214 3	*x 3.2*
	pints, dry	x 51.428	*x 50*
	pints, fluid	x 59.844	*x 60*
	quarts, dry	x 25.714	*x 26*
	quarts, fluid	x 29.922	*x 30*
cubic foot per pound	cubic meters per kilogram	x 0.062 428	*/16 or x 0.06*
cubic foot (Can.)	cubic foot (US)	x 1^	
cubic inch	bushels, struck measure	x 4.650 3 E-04	*x 4.7 E-04*
	cubic feet	/1 728^ or x 5.787 0 E-04	*x 6 E-04*
	cubic meters	x 1.638 7 E-05	*x 1.6 E-05*
	cubic yards	x 2.143 3 E-05	*x 2.1 E-05*
	drams, fluid	x 4.432 9	*x 4.4*
	gallons	/231^ or x 0.004 329 0	*x 0.004 3*
	gills	x 0.138 53	*x 0.14*
	liters	x 0.016 387	*x 0.016*
	milliliters	x 16.387	*x 16*
	minims	x 265.97	*x 270*
	ounces, fluid	x 0.554 11	*x 0.55*
	pecks	x 0.001 860 1	*x 0.001 9*
	pints, dry	x 0.029 762	*x 0.03*
	pints, fluid	x 0.034 632	*x 0.035*
	quarts, dry	x 0.014 881	*x 0.015*
	quarts, fluid	x 0.017 316	*x 0.017*
cubic inch (Can.)	cubic inch (US)	x 1^	
cubic meter	acre-feet	x 8.107 0 E-04	*x 0.11 E-04*
	barrels, petroleum (42 gallons)	x 6.289 8	*x 6.3*
	bushels, struck measure	x 28.378	*x 28*
	cubic feet	x 35.315	*x 35*
	cubic inches	x 61 024	*x 60 000*
	cubic yards	x 1.308 0	*x 1.3 or x 5/4*
	gallons	x 264.17	*x 260*
	liters	x 1 000^	
	milliliters	x 1^E+06	

TABLE 3
Measurement Conversions, By Group

All measurement units are US, unless otherwise noted. All number denominations "billion" and higher are US, unless otherwise noted.

CONVERT	TO EQUIVALENT	BY PRECISELY	OR WITHIN ± 5.0 %
	pecks	x 113.51	*x110*
	pints, dry	x 1 816.2	*x 1 800*
	quarts, dry	x 908.08	*x 900*
	stere	x 1^	
cubic meter per kilogram	cubic feet per pound	x 16.018	*x 16*
cubic rod (Eng.)	cubic feet (Eng.)	x 1 000	
cubic yard	cubic feet	x 27^	
	cubic inches	x 46 656^	*x 47 000*
	cubic meters	x 0.764 55	*x 0.76*
	liters	x 764.55	*x 760*
	milliliters	x 7.645 5 E+05	*x 7.6 E+05*
cubic yard (Can.)	cubic yard (US)	x 1^	
cunit (for solid wood, Can.)	cubic feet	x 100^	
cup, measuring	milliliters	x 236.59	*x 240*
	ounces, fluid	x 8^	
	pints, fluid	/2^ or x 0.5^	
	tablespoons	x 16^	
	teaspoons	x 48^	
cup, measuring (Brit.)	ounces, Imperial fluid (Brit.)	x 10^	
cup, measuring (Can.)	ounces, Imperial fluid (Can.)	x 8^	
dekaliter	gallons	x 2.641 7	*x 2.6*
demiard (Can.)	liters	x 0.284 13	*x 0.28*
demijohn	gallons	x 1 to x 10	
drachm, fluid apothecary (Brit.)	drachm, Imperial fluid (Brit.)	x 1^	
drachm, Imperial fluid (Brit.)	cubic inches	x 0.216 73	*x 0.22*
	drams, fluid (US)	x 0.960 76	*x 1*
	milliliters	x 3.551 6	*x 3.6*
	minims, Imperial (Brit.)	x 60^	
	scruples, Imperial fluid (Brit.)	x 3^	
dram, fluid	cubic feet	x 1.305 5 E-04	*x 1.3 E-04*
	cubic inches	x 0.225 59	*x 0.23*
	gallons	/1 024^ or x 9.765 6 E-04	*x 0.001*
	gills	/32^ or x 0.031 25^	*x 0.03*
	liters	x 0.003 696 7	*x 0.003 7*
	milliliters	x 3.696 7	*x 3.7*
	minims	x 60^	
	ounces, fluid	/8^ or x 0.125^	*x 0.13*
	pints, fluid	/128^ or x 0.007 812 5^	*x 0.008*
	quarts, fluid	/256^ or x 0.003 906 3	*x 0.004*
dram, fluid apothecary	dram, fluid	x 1^	
dram, fluid (Can.)	drachm, Imperial fluid (Brit.)	x 1^	
dram, fluid (US)	drachms, Imperial fluid (Brit.)	x 1.040 8	*x 1*
drop (Can.)	milliliter	/20^ or x 0.05^	
	teaspoon	/100^ or x 0.01^	
fifth (of liquor)	gallons	/5^ or x 0.2^	
	quarts, fluid	x 4/5^ or x 0.8^	
firkin	gallons	x 9	
firkin (Brit.)	Also see MASS		
	barrels, Imperial (Brit.)	/4 or x 0.25 (usually)	

TABLE 3
Measurement Conversions, By Group

All measurement units are US, unless otherwise noted. All number denominations "billion" and higher are US, unless otherwise noted.

CONVERT	TO EQUIVALENT	BY PRECISELY	OR WITHIN ± 5.0 %
	gallons, ale (Brit.)	x 8	
	gallons, Imperial (Brit.)	x 9	
flagon	quarts, fluid	x 2 (usually)	
flask (of mercury)	See MASS		
foot, board	See "board foot"		
foot, mil	See "mil-foot"		
foot, solid	cubic foot	x 1^	
gallon	cubic feet	x 0.133 68	*x 0.13*
	cubic inches	x 231^	*x 230*
	cubic meters	x 0.003 785 4	*x 0.003 8*
	dekaliters	x 0.378 54	*x 0.38*
	drams, fluid	x 1 024^	*x 1 000*
	gallons, dry	x 0.859 37	*x 0.9*
	gallon, fluid	x 1^	
	gallons, Imperial (Brit.)	x 0.832 67	*x 0.8*
	gills	x 32^	
	hectoliters	x 0.037 854	*x 0.038*
	liters	x 3.785 4	*x 3.8*
	milliliters	x 3 785.4	*x 3 800*
	minims	x 61 440^	*x 60 000*
	ounces, fluid	x 128^	*x 130*
	pints, fluid	x 8^	
	quarts, fluid	x 4^	
gallon (Can.)	gallon, Imperial (Brit.)	x 1^	
gallonage	number of gallons	x 1	
gallon, ale (Brit.)	liters	x 4.62	*x 4.6*
gallon, beer (Brit.)	gallon, Imperial ale (Brit.)	x 1^	
gallon, dry (US, not legal)	bushel	x 1/8^ or x 0.125^	*x 0.13*
	cubic inches	x 268.80	*x 270*
	gallons, fluid (US)	x 1.163 6	*x 1.2*
gallon, fluid (Can.)	cubic meters	x 0.004 546 1	*x 0.004 5*
	gallons, fluid (US)	x 1.201 0	*x 1.2*
	gallon, Imperial (Brit.)	x 1^	
gallon, Imperial apothecary (Brit.)	gallon, Imperial (Brit.)	x 1^	
gallon, Imperial (Brit.)	bushels, Imperial (Brit.)	/8^ or x 0.125^	*x 0.13*
	cubic inches	x 277.42	*x 280*
	cubic meters	x 0.004 546 1	*x 0.004 5*
	gallons, fluid (US)	x 1.201 0	*x 1.2*
	liters	x 4.546 1	*x 4.5*
	ounces, Imperial fluid (Brit.)	x 160^	
	pints, Imperial (Brit.)	x 8^	
	quarts, Imperial (Brit.)	x 4^	
gallon, wine (Brit.)	cubic inches	x 231^	*x 230*
gill	cubic feet	x 0.004 177 5	*x 0.004*
	cubic inches	x 7.218 8	*x 7*
	drams, fluid	x 32^	
	gallons	/32^ or x 0.031 25^	*x 0.03*
	liters	x 0.118 29	*x 0.12*
	milliliters	x 118.29	*x 120*
	minims	x 1 920^	*x 2 000*
	ounces, fluid	x 4^	

TABLE 3
Measurement Conversions, By Group

All measurement units are US, unless otherwise noted. All number denominations "billion" and higher are US, unless otherwise noted.

CONVERT	TO EQUIVALENT	BY PRECISELY	OR WITHIN ± 5.0 %
	pints, fluid	/4^ or x 0.25^	
	quarts, fluid	/8^ or x 0.125^	*x 0.13*
gill (Brit.)	cubic meters	x 1.420 7 E-04	*x 1.4 E-04*
gill (Can.)	gill, Imperial (Brit.)	x 1^	
gill, Imperial (Brit.)	pints, Imperial (Brit.)	/4^ or x 0.25^	
hectare-meter	cubic meters	x 10 000^	
hectoliter	bushels, struck measure	x 2.837 8	*x 2.8*
	gallons	x 26.417	*x 26*
hogshead	gallons	x 62.5 to x 140	
	gallons	x 63 (usually)	
	liters	x 238.48	*x 240*
hogshead (Brit.)	gallons (US)	x 64.851	*x 65*
	gallons, Imperial (Brit.)	x 54	
hogshead (Can.)	liters	x 245.49	*x 250*
inch of rain	cubic feet of water per acre	x 3 630^	*x 3 600*
	gallons per square yard	x 5.610 4	*x 5.6*
inch, solid	cubic inch	x 1^	
jeroboam (for wine, Brit.)	bottles (Brit.)	x 4^	
	liters	x 3.0	
kilderkin (Brit.)	barrels, Imperial (Brit.)	/2 or x 0.5	
	gallons, Imperial (Brit.)	x 18	
lambda	cubic millimeter	x 1^	
liter	bushels, struck measure	x 0.028 378	*x 0.28*
	cubic decimeter	x 1^	
	cubic feet	x 0.035 315	*x 0.035*
	cubic inches	x 61.024	*x 60*
	cubic meters	/1 000^ or x 0.001^	
	cubic yards	x 0.001 308 0	*x 0.001 3*
	drams, fluid	x 270.51	*x 270*
	gallons	x 0.264 17	*x 0.26 or x 3/11*
	gallons, Imperial (Brit.)	x 0.219 97	*x 0.22 or x 2/9*
	gills	x 8.453 5	*x 8.5*
	milliliters	x 1 000^	
	minims	x 16 231	*x 16 000*
	ounces, fluid	x 33.814	*x 34*
	pecks	x 0.113 51	*x 0.11*
	pints, dry	x 1.816 2	*x 1.8*
	pints, fluid	x 2.113 4	*x 2.1*
	quarts, dry	x 0.908 08	*x 0.9*
	quarts, fluid	x 1.056 7	*x 1.1*
litre	liter (US)	x 1^	
magnum (for wine, Brit.)	bottles (Brit.)	x 2^	
	liters	x 1.5	
methuselah (for wine, Brit.)	bottles (Brit.)	x 8^	
	liters	x 6.0	
milliliter	cubic centimeter	x 1^	
	cubic feet	x 3.531 5 E-05	*x 3.5 E-05*
	cubic inches	x 0.061 024	*x 0.06*
	cubic meters	x 1^E-06	
	cubic yards	x 1.308 0 E-06	*x 1.3 E-06*
	cups, measuring	x 0.004 226 8	*x 0.004 2*

TABLE 3
Measurement Conversions, By Group

All measurement units are US, unless otherwise noted. All number denominations "billion" and higher are US, unless otherwise noted.

CONVERT	TO EQUIVALENT	BY PRECISELY	OR WITHIN ± 5.0 %
	drams, fluid	x 0.270 51	*x 0.27*
	gallons	x 2.641 7 E-04	*x 2.6 E-04*
	gills	x 0.008 453 5	*x 0.008 5*
	liters	/1 000^ or x 0.001^	
	minims	x 16.231	*x 16*
	ounces, fluid	x 0.033 814	*x 0.034*
	pints, dry	x 0.001 816 2	*x 0.001 8*
	pints, fluid	x 0.002 113 4	*x 0.002 1*
	quarts, fluid	x 0.001 056 7	*x 0.001 1*
mil-foot	one circular mil by one foot	x 1^	
minim	cubic feet	x 2.175 8 E-06	*x 2.2 E-06*
	cubic inches	x 0.003 759 8	*x 0.003 8*
	drams, fluid	/60^ or x 0.016 667	
	drop, fluid (approx.)	x 1	
	gallons	/61 440^ or x 1.627 6 E-05	*x 1.6 E-05*
	gills	/1 920^ or x 5.208 3 E-04	*x 5 E-04*
	liters	x 6.161 2 E-05	*x 6 E-05*
	milliliters	x 0.061 612	*x 0.06*
	ounces, fluid	/480^ or x 0.002 083 3	*x 0.002*
	pints, fluid	/7 680^ or x 1.30 21 E-04	*x 1.3 E-04*
	quarts, fluid	/15 360^ or x 6.510 4 E-05	*x 6.5 E-05*
minim, apothecary	minim	x 1^	
minim, Imperial apothecary (Brit.)	minim, Imperial (Brit.)	x 1^	
minim, Imperial (Brit.)	drachms, Imperial fluid (Brit.)	/60^ or x 0.016 667	*x 0.017*
	scruples, Imperial fluid (Brit.)	/20^ or x 0.05^	
nebuchadnezzar (for wine, Brit.)	bottles (Brit.)	x 20^	
	liters	x 15	
nip (for wine, Brit.)	bottles (Brit.)	/4^ or x 0.25^	
	liters	x 3/16^ or x 0.187 5^	*x 0.19*
ounce, fluid	cubic feet	x 0.001 044 4	*x 0.001*
	cubic inches	x 1.804 7	*x 1.8*
	cubic meters	x 2.957 4 E-05	*x 3 E-05*
	cups, measuring	/8^ or x 0.125^	*x 0.13*
	drams, fluid	x 8^	
	gallons	/128^ or x 0.007 812 5^	*x 0.008*
	gills	/4^ or x 0.25^	
	liters	x 0.029 574	*x 0.03*
	milliliters	x 29.574	*x 30*
	minims	x 480^	*x 500*
	ounces, Imperial fluid (Brit.)	x 1.040 8	*x 1*
	pints, fluid	x 16^ or x 0.062 5^	*x 0.06*
	quarts, fluid	/32^ or x 0.031 25^	*x 0.03*
ounce, fluid apothecary	ounce, fluid	x 1^	
ounce, fluid (Can.)	ounce, Imperial fluid (Brit.)	x 1^	
ounce, Imperial fluid apothecary (Brit.)	ounce, Imperial fluid (Brit.)	x 1^	
ounce, Imperial fluid (Brit.)	cubic meters	x 2.841 3 E-05	*x 2.8 E-05*
	drachms, Imperial fluid (Brit.)	x 8^	
	milliliters	x 28.413	*x 28*
	ounces, fluid (US)	x 0.960 76	*x 1*

TABLE 3
Measurement Conversions, By Group

All measurement units are US, unless otherwise noted. All number denominations "billion" and higher are US, unless otherwise noted.

CONVERT	TO EQUIVALENT	BY PRECISELY	OR WITHIN ± 5.0 %
peck	bushels	/4^ or x 0.25^	
	cubic feet	x 0.311 11	*x 0.31*
	cubic inches	x 537.61	*x 540*
	cubic meters	x 0.008 809 8	*x 0.009*
	liters	x 8.809 8	*x 9*
	pints, dry	x 16^	
	quarts, dry	x 8^	
peck (Can.)	peck, Imperial (Brit.)	x 1^	
peck, Imperial (Brit.)	cubic inches	x 554.84	*x 550*
	gallons, Imperial (Brit.)	x 2^	
	liters	x 9.092 2	*x 9*
perch (for masonry)	cubic feet	x 16.5 or x 24.75 or x 25	
perch (volume)	Also see AREA and LENGTH		
Petrograd standard (for sawed timber, Can.)	cubic feet	x 165	
pint (Can.)	pint, Imperial (Brit.)	x 1^	
pint, dry	bushels	/64^ or x 0.015 625^	*x 0.016*
	cubic feet	x 0.019 445	*x 0.02*
	cubic inches	x 33.600	*x 34*
	cubic meters	x 5.506 1 E-04	*x 5.5 E-04*
	liters	x 0.550 61	*x 0.55*
	milliliters	x 550.61	*x 550*
	pecks	/16^ or x 0.062 5^	*x 0.06*
	quarts, dry	/2^ or x 0.5^	
pint, fluid	cubic feet	x 0.016 710	*x 0.017*
	cubic inches	x 28.875^	*x 29*
	cups	x 2^	
	drams, fluid	x 128^	*x 130*
	gallons	/8^ or x 0.125^	*x 0.13*
	gills	x 4^	
	liters	x 0.473 18	*x 0.47*
	milliliters	x 473.18	*x 470*
	minims	x 7 680^	*x 7 700*
	ounces, fluid	x 16^	
	quarts, fluid	/2^ or x 0.5^	
pint, Imperial apothecary (Brit.)	pint, Imperial (Brit.)	x 1^	
pint, Imperial (Brit.)	gills, Imperial (Brit.)	x 4^	
	ounces, Imperial fluid (Brit.)	x 20^	
pipe	gallons	x 126 (usually)	
	hogsheads	x 2	
	tuns	/2 or x 0.5	
pottle (Brit.)	gallons, Imperial (Brit.)	/2 or x 0.5	
puncheon	gallons	x 84 (usually)	
puncheon (Brit.)	barrels, Imperial (Brit.)	x 2	
	gallons, Imperial (Brit.)	x 72 (usually)	
quart (Can.)	quart, Imperial (Brit.)	x 1^	
quarter, Imperial (volume, Brit.)	bushels, Imperial (Brit.)	x 8^	
quart, dry	bushels, struck measure	/32^ or x 0.031 25^	*x 0.03*
	cubic feet	x 0.038 889	*x 0.04*
	cubic inches	x 67.201	*x 67*
	cubic meters	x 0.001 101 2	*x 0.001 1*

TABLE 3
Measurement Conversions, By Group

All measurement units are US, unless otherwise noted. All number denominations "billion" and higher are US, unless otherwise noted.

CONVERT	TO EQUIVALENT	BY PRECISELY	OR WITHIN ± 5.0 %
	liters	x 1.101 2	*x 1.1*
	pecks	/8^ or x 0.125^	*x 0.13*
	pints, dry	x 2^	
quart, dry (US)	quarts, Imperial (Brit.)	x 0.968 94	*x 0.97*
quart, fluid	cubic feet	x 0.033 420	*x 0.033*
	cubic inches	x 57.75^	*x 60*
	drams, fluid	x 256^	*x 260*
	gallons	/4^ or x 0.25^	
	gills	x 8^	
	liters	x 0.946 35	*x 0.95*
	milliliters	x 946.35	*x 950*
	minims	x 15 360^	*x 15 000*
	ounces, fluid	x 32^	
	pints, fluid	x 2^	
	quarts, dry	x 0.859 37	*x 0.9*
quart, fluid (US)	quarts, Imperial (Brit.)	x 0.832 67	*x 0.8*
quart, Imperial (Brit.)	ounces, Imperial fluid (Brit.)	x 40^	
	pints, Imperial (Brit.)	x 2^	
	quarts, dry (US)	x 1.032 1	*x 1*
	quarts, fluid (US)	x 1.200 9	*x 1.2*
rain, inch of	See "inch of rain"		
rehoboam (for wine, Brit.)	bottles (Brit.)	x 6^	
	liters	x 4.5	
rod	Also see AREA and LENGTH		
rod, cubic (Eng.)	See "cubic rod"		
salmanazar (for wine, Brit.)	bottles (Brit.)	x 12^	
	liters	x 9.0	
salt cart (Can.)	liters	x 490.98	*x 500*
salt tub (Can.)	salt cart (Can.)	/6^ or x 0.166 67	
scruple, Imperial fluid (Brit.)	drachms, Imperial fluid (Brit.)	/3^ or x 0.333 33	*x 0.33*
	minims, Imperial (Brit.)	x 20^	
	ounce, Imperial fluid (Brit.)	/24^ or x 0.041 667	
specific volume	1/density	x 1^	
	cubic meter per kilogram	x 1^	
split (for beverage containers)	ounces, fluid	x 6 (usually) [20]	
stere	cubic meter	x 1^	
	kiloliter	x 1^	
tablespoon, measuring	cup	/16^ or x 0.062 5^	*x 0.06*
	drams, fluid	x 4^	
	milliliters	x 14.787	*x 15*
	ounces, fluid	/2^ or x 0.5^	
	pints, liquid	/32^ or x 0.031 25^	
	teaspoons	x 3^	
tablespoon, measuring (Brit.)	ounces, Imperial fluid (Brit.)	x 5/8^ or x 0.625^	*x 15*
tablespoon, measuring (Can.)	ounces, Imperial fluid (Can.)	/2^ or x 0.5^	
teaspoon, measuring	drams, fluid	x 4/3^ or x 1.333 33	*x 1.3*
	milliliters	x 4.928 9	

[20] A "split" is half the usual volume of a beverage.

Measurement Conversions, By Group

All measurement units are US, unless otherwise noted. All number denominations "billion" and higher are US, unless otherwise noted.

CONVERT	TO EQUIVALENT	BY PRECISELY	OR WITHIN ± 5.0 %
	ounces, fluid	/6^ or x 0.166 67	*x 0.67*
	tablespoons	/3^ or x 0.333 33	*x 0.33*
teaspoon, measuring (Brit.)	ounces, Imperial fluid (Brit.)	x 5/24^ or x 0.208 33	
teaspoon, measuring (Can.)	ounces, Imperial fluid (Can.)	/6^ or x 0.166 67	
ton (volume)	Also see ENERGY, FORCE, and MASS		
tonnage	Also see MASS		
tonnage, gross register (for ships)	tonnage, gross	x 1^	
tonnage, gross (for ships)	total register-ton capacity less nation-defined spaces	x 1	
tonnage, net register (for ships)	tonnage, net	x 1^	
tonnage, net (for ships)	gross tonnage less nation-defined spaces unavailable for cargo	x 1	
tonnage, register under-deck	number of register tons capacity under tonnage deck	x 1	
tonnage, ship	tonnage, gross (usually)	x 1	
tonnage, vessel	See a specific other "tonnage"		
ton, displacement (for ships)	cubic feet of fresh water	x 35.9	*x 36*
	cubic feet of sea water	x 35	
ton, English water (Brit.)	gallons (US)	x 270.91	*x 270*
	gallons, Imperial (Brit.)	x 224^	*x 220*
ton, freight	ton, measurement	x 1^	
ton, measurement (for ships)	cubic feet	x 40^	
ton, register	cubic feet	x 100^	
	cubic meters	x 2.831 7	*x 2.8*
ton, shipping	ton, measurement	x 1^	
	bushels	x 32.143	*x 32*
	cubic feet	x 40^	
ton, shipping (Brit.)	bushels (US)	x 33.750	*x 33*
	bushels, Imperial (Brit.)	x 32.701	*x 33*
	cubic feet	x 42^	
ton, shipping (US)	bushels, Imperial (Brit.)	x 31.144	*x 30*
ton, water (Brit.)	gallons (US)	x 270.91	*x 270*
	gallons, Imperial (Brit.)	x 224^	*x 220*
tub (for herring, Can.)	herring barrel	/2 or x 0.5	
tun (volume)	Also see TIME		
	gallons	x 252 (usually)	
	hogsheads	x 4	
tun (volume, for ships' water, Brit.)	gallons, Imperial (Brit.)	x 210^	
tunnage	tonnage	x 1^	
yard of ale	pints, fluid	x 2 or x 3	
VOLUME FLOW (see FLOW RATE)			
VOLUME PER UNIT TIME (see FLOW RATE)			
WEIGHT (see MASS and FORCE)			
WORK (see ENERGY)			

EARTHQUAKES

Modified Mercalli Intensity (MM) Scale:

The MM scale expresses the intensity of an earthquake according to the effects at a given location.

Level	Effects
I	Not felt by people except for a very few under especially favorable conditions. Registered by seismographs.
II	Very Weak: Felt by a few people at rest, especially on upper floors of buildings. Delicately suspended objects may swing.
III	Weak: Felt noticeably by many people indoors, especially on upper floors of buildings. Many people do not recognize the earthquake as such; vibration is similar to that from a passing truck. Standing vehicles may rock. Duration can be estimated.
IV	Not Strong: Felt indoors by many, outdoors by few during the day. Some people awakened at night. Dishes, windows, doors disturbed; walls make cracking sound. Sensation as from a heavy truck striking building. Standing vehicles rock noticeably.
V	Fairly Strong: Felt by nearly everyone, many awakened. Some dishes, windows broken. Unstable objects overturned. Pendulum clocks may stop.
VI	Strong: Felt by all people, many frightened. Some heavy furniture moved; a few instances of fallen plaster. Damage slight.
VII	Very Strong: Damage negligible in well designed and well built buildings, slight to moderate in well built ordinary structures, considerable in badly built designed or built structures. Some chimneys broken.
VIII	Damaging: Damage slight in specially designed structures, considerable and with partial collapse in ordinary substantial buildings, great in poorly built structures. Chimneys, factory stacks, columns, walls, monuments fall. Heavy furniture overturned.
IX	Destructive: Considerable in specially designed structures, great and with partial collapse in substantial buildings. Well designed frame structures thrown out of plumb. Buildings shifted off foundations.
X	Very Destructive: Some well built wooden structures destroyed, most masonry and frame structures destroyed with foundations. Rails bent.
XI	Catastrophic: Few, if any, masonry structures remain standing, bridges destroyed. Rails bent greatly.
XII	Exceptionally Catastrophic: Damage total. Lines of sight and level distorted. Objects thrown into the air.

Richter Scale (for earthquakes):

The original Charles F. Richter earthquake scale, introduced in 1935, expresses the magnitude of an earthquake in terms of ground movement and energy released at the place where the earthquake occurred. The magnitude is derived from seismographic measurements of seismic-wave amplitude. Freed energy bears an empirical relationship to the magnitude, as defined by B. Gutenberg and Richter in the 1950s.

A "seismic-moment scale", or "moment-magnitude scale", was proposed by Hiroo Kanamori in 1977. As compared with the original Richter scale, the seismic-moment scale provides a more accurate measurement of magnitude for very large earthquakes, magnitude 7, or greater. The seismic moment is developed from the average displacement, the rupture area of the earth fault, and the rigidity of the rock.

The seismic-moment scale has been adapted to match the Richter scale for freed energy so that both change mathematically in the same way with respect to changes of magnitude. The analysis required to determine the seismic-moment is relatively slow, with correspondingly slow reporting of results. (For effects of earthquake magnitude changes, see "Richter scale" in the alphabetical Table 2 or in Table 3 under ENERGY: LENGTH; or TORQUE.)

ELECTROMAGNETIC SPECTRUM

The types of radiation listed below are not necessarily exclusive, and their values are nominal.

Frequency, Hertz	Wave Length, Meters	Type of Radiation
1 E+23*	3 E-15*	Cosmic photons
1 E+22	3 E-14	Gamma rays
1 E+21	3 E-13	Gamma rays and X-rays
1 E+20	3 E-12	X-rays
1 E+19	3 E-11	Soft X-rays
1 E+18	3 E-10	X-rays and ultraviolet
1 E+17	3 E-09	Ultraviolet
1 E+16	3 E-08	Ultraviolet
1 E+15	3 E-07	Visible light
1 E+14	3 E-06	Infrared
1 E+13	3 E-05	Infrared
1 E+12	3 E-04	Far infrared
1 E+11	0.003	Microwaves
1 E+10	0.03	Microwaves, radar
1 E+09	0.3	Radar
1 E+08	3	Television and FM radio
1 E+07	30	Shortwave radio
1 E+06	300	AM radio
1 E+05	3 000	Longwave radio
1 E+04	3 E+04	Induction heating
1 000	3 E+05	Electronic devices
100	3 E+06	Power
10	3 E+07	Power
1	3 E+08	Commutated direct current
0	Infinite	Direct current

* For equivalent powers of 10, see Table 1.

NUMBERS

Confidence Level (statistical):

Standard Deviation, Sigma	Confidence Level, Percent
0.674 4	50.00
0.755 3	55.00
0.841 5	60.00
0.934 6	65.00
1.000	68.27
1.037	70.00
1.150	75.00
1.282	80.00
1.439	85.00
1.645	90.00
1.960	95.00
2.00	95.44
2.05	96.00
2.17	97.00
2.33	98.00
2.58	99.00
3.00	99.73
3.29	99.90
3.9	99.99

Relative Values of Number Denominations:

Each of the US number denominations above "thousand" has a magnitude 1 000 times greater than that of the preceding denomination. Thus, after "thousand," there are "million," "billion," "trillion," "quadrillion," "quintillion," etc.

The British (and other nations') number denominations above "thousand" begin with a magnitude 1 000 times that of the preceding denomination. Thus, after "thousand," there are "million," "milliard," and "billion." But above "billion," each of the denominations "trillion," "quadrillion," "quintillion," etc., has a magnitude one million (1 000 000) times that of the preceding denomination.

Roman Numbers:

Roman Numbers	Arabic Numbers
I	1
V	5
X	10
L	50
C	100
D	500
M	1 000
$\overline{V}$	5 000
$\overline{X}$	10 000
$\overline{L}$	50 000
$\overline{C}$	100 000
$\overline{D}$	500 000
$\overline{M}$	1 000 000

Roman numerals provide other numbers by the following rules:

1. A bar at the top of a symbol signifies a multiplier of one thousand, as shown in the above list.
2. A symbol following another of equal or greater value adds its value, e.g.: II = 1 + 1 = 2, VI = 5 + 1 = 6.
3. A symbol preceding another of greater value subtracts its value, e.g.: IV = 5 - 1 = 4, XC = 100 - 10 = 90.
4. A symbol that is between two symbols of greater value subtracts its value from that of the second symbol, and the result is added to the value of the first symbol: XXIV = 20 + (5 - 1) = 24, DVLI = 500 + (50 - 5) + 1 = 546.
5. Short forms of numbers to be subtracted are preferred to long forms of numbers to be added, e. g., MCM rather than MDCCCC for the year 1900.
6. The first symbol of a group should be the symbol having the highest value; then that symbol's subtrahend, if any; then other symbols descending in sequence toward the lowest-value, e.g.; XIV, not VIX, = 14; XCVII = 97.

TEMPERATURE COLOR SCALE (for iron and steel)

Color	Temperature (approximate)	
	Deg. F	Deg. C
Black	not above 900	not above 480
Faint red	900 - 1050	480 - 570
Blood red	1050 - 1170	570 - 630
Dark cherry red	1170 - 1240	630 - 670
Medium cherry red	1240 - 1370	670 - 740
Full cherry red	1370 - 1540	740 - 840
Bright red	1540 - 1650	840 - 900
Red orange	1650 - 1730	900 - 940
Orange	1730 - 1810	940 - 990
Lemon yellow	1810 - 1980	990 - 1080
Light yellow	1980 - 2190	1080 - 1200
White	2190 and above	1200 and above

TIME

Geologic Time Scale (approximate):

Era	Period	Epoch	Years Ago
Cenozoic	Quaternary	Holocene (recent)	11 000
		Pleistocene (glacial)	500 000 - 2 000 000
	Tertiary	Pliocene	13 000 000
		Miocene	25 000 000
		Oligocene	36 000 000
		Eocene	58 000 000
		Paleocene	63 000 000
Mesozoic	Cretaceous		135 000 000
	Jurassic		180 000 000
	Triassic		230 000 000
Paleozoic	Permian		280 000 000
	Pennsylvanian (Upper Carboniferous)		310 000 000
	Mississippian (Lower Carboniferous)		345 000 000
	Devonian		405 000 000
	Silurian		425 000 000
	Ordovician		500 000 000
	Cambrian		600 000 000
Proterozoic Precambrian			more than 600 000 000
Archeozoic Precambrian			

Time of Day:

12-Hour Clock	24-Hour Clock	Number of Ship's Bells
12:30 A.M.	0030	1
1:00 A.M.	0100	2
1:30 A.M.	0130	3
2:00 A.M.	0200	4
2:30 A.M.	0230	5
3:00 A.M.	0300	6
3:30 A.M.	0330	7
4:00 A.M.	0400	8
4:30 A.M.	0430	1
5:00 A.M.	0500	2
5:30 A.M.	0530	3
6:00 A.M.	0600	4
6:30 A.M.	0630	5
7:00 A.M.	0700	6
7:30 A.M.	0730	7
8:00 A.M.	0800	8
8:30 A.M.	0830	1
9:00 A.M.	0900	2
9:30 A.M.	0930	3
10:00 A.M.	1000	4
10:30 A.M.	1030	5
11:00 A.M.	1100	6
11:30 A.M.	1130	7
12:00 A.M.	1200	8
12:30 P.M.	1230	1
1:00 P.M.	1300	2
1:30 P.M.	1330	3
2:00 P.M.	1400	4
2:30 P.M.	1430	5
3:00 P.M.	1500	6
3:30 P.M.	1530	7
4:00 P.M.	1600	8
4:30 P.M.	1630	1
5:00 P.M.	1700	2
5:30 P.M.	1730	3
6:00 P.M.	1800	4
6:30 P.M.	1830	5
7:00 P.M.	1900	6
7:30 P.M.	1930	7
8:00 P.M.	2000	8
8:30 P.M.	2030	1
9:00 P.M.	2100	2
9:30 P.M.	2130	3
10:00 P.M.	2200	4
10:30 P.M.	2230	5
11:00 P.M.	2300	6
11:30 P.M.	2330	7
12:00 P.M.	2400	8

WIND SPEEDS

Beaufort Scale:

Beaufort Force	Wind Strength	Wind Speed, per hour	
		Miles	Kilometers
0	calm	0 - 1	0 - 1.6
1	light air	1 - 3	1.6 - 4.8
2	light breeze	4 - 7	6.4 - 11.3
3	gentle breeze	8 - 12	12.9 - 19.3
4	moderate breeze	13 - 18	20.9 - 29.0
5	fresh breeze	19 - 24	30.6 - 38.6
6	strong breeze	25 - 31	40.2 - 49.9
7	moderate gale	32 - 38	51.5 - 61.2
8	fresh gale	39 - 46	62.8 - 74.0
9	strong gale	47 - 54	75.6 - 86.9
10	whole gale	55 - 63	88.5 - 101.4
11	storm	64 - 72	103.0 - 115.9
12	hurricane	73 or more	117.5 or more

FPP Tornado Scale:

FPP Scale	Wind Speed, per hour	
	Miles	Kilometers
- - -	less then 40	less than 64
0	40 - 72	64 - 116
1	73 - 112	118 - 180
2	113 - 157	182 - 253
3	158 - 206	254 - 332
4	207 - 260	333 - 418
5	261 - 318	420 - 512
6	- - -	- - -
7	- - -	- - -

Fujita Intensity Scale (F-scale):

F-scale	Damage	Wind Speed, per hour	
		Miles	Kilometers
F0	light	less than 72.1	less than 116
F1	moderate	72.1 - 111.8	116 - 180
F2	considerable	112.5 - 157.2	181 - 253
F3	severe	157.8 - 206.3	254 - 332
F4	devastating	206.9 - 260.4	333 - 419
F5	incredible	more than 260.4	more than 419

Saffir-Simpson Hurricane Scale (SSH) (does not apply to the Pacific Islands):

Hurricane Category	Wind Strength	Wind Speed, per hour	
		Miles	Kilometers
1	weak	75 - 95	120.7 - 152.9
2	moderate	96 - 110	154.4 - 177.0
3	strong	111 - 130	178.6 - 209.2
4	very strong	131 - 155	210.8 - 249.2
5	devastating	more than 155	more than 249.2

Tropical Wind Scale:

	Wind Speed, per hour	
	Miles	Kilometers
Tropical Depression	not above 39	not above 62.8
Tropical Storm	39 - 73	62.8 - 117.5